DEUTSCHE FORSCHUNGS- UND VERSUCHSANSTALT
FÜR LUFT- UND RAUMFAHRT E.V. (DFVLR)

SOLAR THERMAL ENERGY UTILIZATION

German Studies on Technology and Application

Editor: M. Becker

Volume 2:
Technologies of Heat Exchangers
(Receiver/Reformer) and Storage

Springer-Verlag
Berlin Heidelberg GmbH 1987

Dr.-Ing. Manfred Becker

Hauptabteilung Energietechnik der
Deutschen Forschungs- und Versuchsanstalt für Luft- und Raumfahrt e. V. (DFVLR), Köln

ISBN 978-3-540-18031-9   ISBN 978-3-662-01628-2 (eBook)
DOI 10.1007/978-3-662-01628-2

2362/3020-543210

Preface

The energy crisis in 1973 and 1979 initiated a great number of activities and programs for low and high temperature application of solar energy. Synthetic fuels and chemicals produced by solar energy is one of them, where temperatures in the range of $600-1000\,^{\circ}C$ or even higher are needed. In principle such high temperatures can be produced in solar towers. For electricity production, the feasibility and operation of solar tower plants has been examined during the SSPS - project (Small Solar Power System) in Almeria, Spain.

The objective of Solar Thermal Energy Utilization is to extend the experience from the former SSPS - program in to the field of solar produced synthetic fuels. New materials and technologies have to be developed in order to research this goal. Metallic components now in use for solar receivers need to be improved with respect to transient operation or possibly replaced by ceramics. High temperature processes, like steam-methane reforming, coal conversion and hydrogen production need to be developed or at least adapted for the unconventional solar operation. Therefore Solar Thermal Energy Utilization is a long term program, which needs time for its development much more time than the intervals expected in between further energy crisis. The "Studies on Technology and Application on Solar Energy Utilization" is a necessary step in the right direction in order to prepare for the energy problems in the future.

Prof. Dr. H. F. Knoche

Rheinisch-Westfälische Technische Hochschule Aachen, Federal Republic of Germany

SOLAR THERMAL ENERGY UTILIZATION

German Studies on Technology and Application

**Volume 2:. Technologies of Heat Exchangers
(Receiver/Reformer) and Storage**

Contents

Page

Volumetric Ceramic Receiver Cooled by Open Air Flow   1
- Feasibility Study -
K. Freudenstein, B. Karnowsky,
Interatom, Bergisch-Gladbach

A Multistage Steam Reformer Utilizing Solar Heat   57
W. Jäger, U. Leuchs, W. Siebert,
Interatom, Bergisch-Gladbach

Layout of High Temperature Solid Heat Storages   111
H. Kalfa, Chr. Streuber,
Didier, Wiesbaden

Expert Opinion and Co-operation in the Development   211
Program High Temperature Storage Tank
Th.J. Bohn, K. Werner, W. Bitterlich, F.J. Josfeld,
Uni-Essen GHS

Index of Authors   319

Volume 1: General Investigations on Energy Availability

Yearly Yield of Solar CRS-Process Heat and Temperature
of Reaction,
P. Koepke, H. Quenzel, R. Sizmann,
Universität-München

Literature Survey in the Field of Primary and Secondary
Concentrating Solar Energy Systems Concerning the Choice
and Manufacturing Process of Suitable Materials,
A. Grychta, J. Kaufmann, P. Lippert, G. Lensch,
NU-Tech, Neumünster

Considerations and Proposals for Future Research and
Development of High Temperature Solar Processes,
F. Boese, P.E. Huber, H.W. Kappler, J. Lammers,
Motor Columbus, Stuttgart

Volume 3: Solar Thermal Energy for Chemical Processes

Steam Reforming of Methane Utilizing Solar Heat
W.D. Müller, Lurgi, Frankfurt

Solar Steam Reforming of Methane (SSRM) Program Proposals,
A. Kalt, DFVLR, Köln

Solar Steam Reforming of Methane - Program Proposals
U. Leuchs, Interatom, Bergisch-Gladbach

Comparative Investigations and Ratings of Different Solar
Systems Using Tubular Steam Reformers
W. D. Müller, Lurgi, Frankfurt and H. Fuhrmann,
MAN-Technologie GmbH, München

Process Synthesis of a Gasification Process Modified for
High Solar Energy Integration,
G. Birke, R. Reimert, Lurgi, Frankfurt

Utilization of Solar Energy for Hydrogen Production by
High Temperature Electrolysis of Steam,
E. Erdle, J. Groß, V. Meyringer,
Dornier, Friedrichshafen

INTERATOM

VOLUMETRIC CERAMIC RECEIVER
COOLED BY OPEN AIR FLOW
- FEASIBILITY STUDY -

K. FREUDENSTEIN
B. KARNOWSKY

INTERATOM, BERGISCH-GLADBACH

Symbolics

L	Channel Length / m /	
s	Channel width / m /	
	= hydraulic diameter of the channel	
b	Wall thickness between two adjacent channels / m /	
x	Distance along the channel / m /	
Nu	Nusselt number	
Re	Reynolds number	
Pe	Peclet number	
w	Gas velocity / m/s /	
ν	Air viscosity / m^2/s /	
λ	Conductivity / W/mK /	
$A = s^2$	Cross section area of orifice / m^2 /	
$A_F = (s + b)^2$	Face (back side) area with orifice / m^2 /	
ζ	Pressure loss coefficient	
ρ	Density / kg/m^3 /	
Δp	Pressure loss / Pa /	
$\dot{m}$	Flow rate / kg/s /	
η	Efficiency	
$\dot{q}$	Heat flux / W/m^2 /	
ε	Emissivity	

Indices

F	Face	in	inlet
A	Air	out	outlet
S	Solid (ceramic)	M	mean
C	Channel	St	"Standard"
f	friction	B	Backscattering, reflection
l	loss		
R	Radiation	T	Thermal

Contents

1   Introduction   61

2   Steam Reforming of Methane   65

3   Steam Reforming Plants   67

3.1 Conventional Plants   67
3.2 Convectionally Heated Steam Reformers   70
3.3 Multi-Stage Steam Reformer with Convectional Heating   73

4   Steam Reforming Plants with Solar Heating   78

4.1 Determination of the Design Data   78
4.2 Description of the Plant   81
4.3 Design of the Steam Reformer   88
4.4 Necessary Development Work   98
4.5 Characteristics of Operation   101
4.6 Costs of Manufacturing the Reformer   103

5   Literature   104

6   Figures   105

1   Introduction and Task Description

To find out whether a volumetric ceramic receiver could
be suitable for high temperature application in an
open air cooling system, a feasibility study with
respect to thermal loading was commissioned by DFVLR *)
to Interatom GmbH. The ceramic type receiver should
be orientated at manufacture possibilities of Silicon
Carbide and Silicon Nitride ceramic heat exchangers
of HOECHST Ceram Tec.

The solar flux on the receiver was defined to be homogeneously distributed and limited to 100 W/cm^2. By
thermodynamical analysis it has to be shown for
steady-state conditions that the thermal efficiency
of the receiver could be optimised by reducing the
radiation losses and increasing the heat transfer to
the cooling gas by variation of the channel geometries
within the manufacturer limits.

Besides these feasibility tests considerations on
pressure drop and directional distribution of solar
flux are to be taken into account.

For this purpose a heat transfer model for the receiver
has to be developed and used for the parameter tests
taking into account radiative, convective and con-
ductive heat transfer and its interchange. The method
to be used to calculate the radiative interchange is
on principle an exact one, because it is determined
by Monte-Carlo methods.

*) Deutsche Forschungs- und Versuchsanstalt für Luft- und
    Raumfahrt

2   Description of Model

         To calculate the temperature distribution in the solid
         part of the receiver and in the air flow a general
         purpose heat transfer code LIWAK and ALBEMO, a code
         for calculation of radiation heat transfer coefficients,
         were applied.

2.1   Short Codes Description (LIWAK, ALBEMO)

         The overall heat exchange within the model was cal-
         culated by numerical treatment using the code LIWAK / 1 /.
         The code calculates the temperatures of a system of
         nodes in any configuration. Heat may be exchanged
         in any appropriate mode (conduction, convection,
         radiation) between any desireable number of nodes
         without any geometric restrictions. The code requires
         the input of details about each node regarding type
         (normal, flow, boundary), heat capacity, flow charac-
         teristics etc. and about each heat transfer coefficient.
         A user supplied routine may be linked to the code
         to set any node parameter (temperature heat transfer
         coefficient, flow data etc.). Transient as well as
         steady state conditions may be computed.

         The code ALBEMO / 2 / calculates transport factors
         to be used to describe heat transfer due to thermal
         radiation between the surface elements of a rectangular
         cavity. These transport factors are determined by the
         geometry of the cavity and by the reflection and
         absorption properties of its boundaries. They are
         defined as the product of the size of the emitting
         element, its emittance and the probability that a
         quantum starting on one surface will be absorbed by
         another surface element.

The calculation is based on the Monte Carlo method. Several reflections at the walls of the cavity could take place between emission and absorption. The code has various options to describe the angular emittance and absorptance. The reflection is assumed to consist of a specular or a diffuse component with the choice of optional portions.

The transport factors computed by ALBEMO are used as input data for general heat transport programs such as LIWAK to calculate temperature distributions.

2.2   Description of LIWAK Model

The input data in LIWAK were fed in by precalculated heat transfer coefficients between the nodes, some of which were appropriately transformed in the user's subroutine during the iteration procedure of the code (see below):

-    for conduction along the channel wall of the solid ceramics;

-    for convection between the solid wall nodes and the gas flow section;

-    for radiative exchange between the nodes and the heat source and the nodes.

For the latter two modes it is assumed that the solid node temperature of the wall could be taken as well as the wall surface temperature of the channel section because of the comparatively low thermal resistance in the small solid wall.

The radiative exchange factors were provided by ALBEMO calculations with the same node distribution as for the LIWAK model and for diffuse reflection.

The geometrical model, the formulas of convective heat transfer coefficients and the formulas of pressure loss were fed in the user's subroutine as described below.

2.2.1 Channel Geometry

An idealised model of a representative receiver channel was chosen to define the nodes distribution for the numerical heat transfer calculation.

Generally a square-type orifice of the channel was taken. The ratio L/s of channel length L to width s was taken L/s = 10 initally and varied for prolongation of the channel. The section lengths along the channel were always chosen in a fixed ratio to the channel length, with the relative shortest sections at the channel inlet increasing to the back of the channel. The sequence of ratios $\Delta x/L$ introduced for totally ten sections was always:
.01, .01, .02, .02, .04, .05, .1, .15, .20, .40.

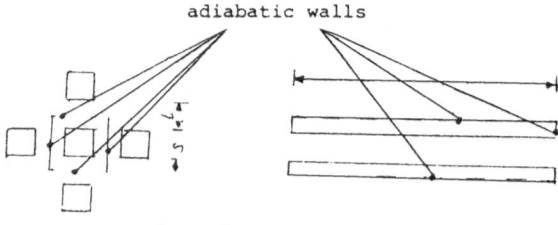

adiabatic walls

channel

cross section   axial cut

Abiabatic boundary conditions were taken in the middle
of the channel wall between adjacent channels. This
was true for conduction at the back end of the channel
wall too.

2.2.2   Convective Heat Transfer

Convective heat exchange to the gas flow was assumed
at the face area and in the channel, but not at
its back side.

Correlations for empirical heat transfer coefficients
at the face area with orifices are scarcely to find
in the literature.

Finally, a correlation from tubes in cross flow by
Churchill and Bernstein / 3 / that covers the
complete range of available data was chosen:

$$Nu_F = 0.3 + \frac{0.62 \, Re^{1/2} \, Pr^{1/3}}{\left[1 + (\frac{0.4}{Pr})^{2/3}\right]^{3/4}} \left[1 + (\frac{Re}{282000})^{5/8}\right]^{4/5}$$

valid for $10^2 < Re < 10^7$; $Pe > 0.2$.

The wall thickness b of the channel was used as the
characteristic length in the Reynolds number

$$Re_F = \frac{W_F \cdot b}{\nu}$$

and in the heat transfer coefficient correlation:

$$\alpha_F = Nu_F \cdot \frac{\lambda}{b} \cdot$$

The flow in the channel was considered as developing flow, which means slug flow at the entrance changing into laminar flow, at uniform wall temperature. For this case Kays / 4 / derived the following empirical correlation:

$$Nu_C = 3.66 + \frac{0.104 \cdot s/x \cdot Re_C \cdot Pr}{1 + 0.016 \left[s/x\ Re_C\ Pr \right]^{0.8}}$$

in which the hydraulic diameter of a square cross-section channel s and the distance along the channel x has been introduced already:

$$Re_C = \frac{w_C \cdot s}{\nu}$$

$$\alpha_C = Nu_C \cdot \frac{\lambda}{s}$$

To account for temperature dependent physical properties of the heated air a correction factor is applied on the preceeding Nusselt number correlations / 5 /:

$$\left(\frac{T}{T_w}\right)^{0.45} \qquad\qquad 0.5 < \frac{T}{T_w} < 1.5$$

the temperatures being T/K, the gas temperature, and T_w/K, the wall temperature in corresponding sections of gas flow and channel wall.

2.2.3 Pressure Loss

The pressure loss for the gas flow, passing through the receiver channel, is calculated from

$$\Delta p = (\zeta_1 + \zeta_f) \frac{\rho}{2} w^2$$

taking into account the entrance, exit and friction losses / 6 /.

For $\quad w = \dfrac{\dot{m}}{\rho \cdot A}$

the formula changes in

$$\Delta p = (\zeta_1 + \zeta_f) \frac{1}{2\rho} (\frac{\dot{m}}{A})^2$$

i. e.

$$\Delta p = (\overline{\zeta}_1 + \overline{\zeta}_f) \frac{1}{2} (\frac{\dot{m}}{A})^2$$

For the entrance and exit losses sharp edged con-
tractions or enlargements and velocity heads
are assumed, therefore the pressure loss coefficients
were taken as follows / 7 /:

at the inlet $\quad \overline{\zeta}_{1_{in}} = \left[1.5 - (\frac{A}{A_F})^2 \right] \frac{1}{\rho_{in}}$

at the outlet $\quad \overline{\zeta}_{1_{out}} = \left[2 \frac{A}{A_F} (\frac{A}{A_F} - 1) \right] \frac{1}{\rho_{out}}$

The friction losses for laminar flow in a rectangular channel are according to / 7 /:

$$\bar{\zeta}_f = \zeta_f \cdot \frac{1}{\rho_M} = \frac{64}{1.13 \; Re} \frac{L}{s} \frac{1}{\rho_M} \; .$$

The incident solar flux was taken constant and homogeneously distributed with respect to its power of 100 W/cm^2 throughout the parameter studies. This value was assumed to be already reduced by direct reflection losses at the front plate (to be discussed in Chapter 4.1). This value is approximately the maximum solar flux to be achieved with the parabolic mirror field at Lampoldshausen.

To get a feeling for the most sensitive parameters some parameters were fixed at reasonable values, while others were varied in three pretest calculations with the analysis model:

- the geometry of the cross-section orifice,

- the physical properties,

- the radiation distribution and the incident angle.

The fixed parameters were chosen as follows:

- the mass flow rate of the air was taken as $\frac{\dot{m}}{A_F} = \dot{m}_{St} = 0.934$ $kg/m^2 s$ related to the cross-section of the orifice and half the wall thickness around the orifice of the channel. This value was chosen such as to heat up the air from 20 °C inlet to 1000 °C maximum outlet temperature, i. e. by 980 K, if all incident radiation would have been overtaken by the gas flow without any radiative losses.

	Thermal Conductivity W/mK	Emissivity	Convective heat transfer coefficient W/m²K	Radiation distribution	Channel width mm
Pretest 1 "Geometry"	50	0.8	See Chapter 2.2.2	Cosine	Variation 1, 3, 5 or 10
Pretest 2 "Physical Properties"	Variation 25 or 100	0.6	50 % or 200 % of α-formula in Chapter 2.2.2	Cosine	3
Pretest 3 "Radiation Distribution"	50	0.8	See Chapter 2.2.2	Variation parallel, inclination angle to channel axis 2.9°, 5.7° or 30°	3

Table 3.1   Parameter Sets for Pretests

- the ratio ten for channel length to width
  L/s = 10

- the channel wall thickness, i. e. the distance
  between two adjacent channels, b = 3 mm,
  the lowest manufacturer limit.

3.1   Data Sets for Pretests
       (Table 3.1)

For the first pretest the square cross-section length
of the orifice was varied from 1 mm, 3 mm, 5 mm
to 10 mm.

For the subsequent pretests the "best" one (see
Chapter 4.1) from the geometry test was chosen for
further analyses.

For the second pretest the physical properties were
varied to get an idea of their sensitivity to the
temperature distribution and the thermal efficiency,
to account for uncertainties in their determination
and as far as the relative low conductivity of
Silicon nitride is concerned, to look for the
appropriate material.

In the third pretest the cosine distribution of the
incident radiation was changed into parallel radiation,
having different inclination angles to the channel
axis, as shown in the following sketch,

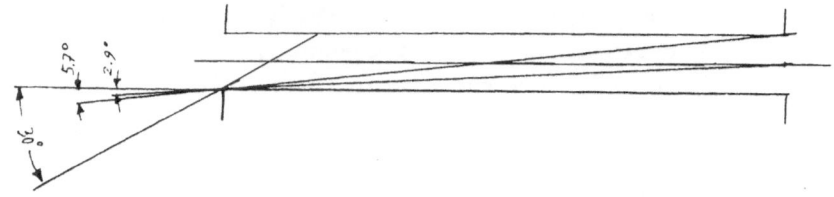

i. e. the two smaller angles are chosen such as to
reach the back end corner of the channel coming
in from the centre-line of the channel or the face
edge respectively. In effect a small part of incident
radiation would be emitted through the back side
orifice and lost for the thermal interchange with
the gas flow in these cases.

3.2   Data Sets for Feasibility Test
        (Table 3.2)

The feasibility test was undertaken with the knowledge
of the pretest results and therefore limited to few
parameter variations, especially

-       the channel length, i. e. the ratio L/s of
        channel length to width,

-       the mass flow rate.

To reach the goal of feasibility, i. e. gas outlet
temperatures of 1000 °C with respecting less 1400 °C
maximum ceramic wall temperatures (near melting
point of Silicon in SiSiC), the most promising channel
geometries of pretest 1 were taken, with channel width
of 1 or 3 mm. The cosine distribution for incident
radiation, and for the physical properties the most
probable ones from pretest 1 and 3 have been retained.

Besides one test was undertaken with a larger wall
thickness of the channel, to analyse the importance
of heat conduction in the wall.

Channel width s/mm	Relative channel length L/s	Relative mass flow rate $\dot{m}_{A_F}/\dot{m}_{St}$	Wall thickness b/mm
1	10	1	3
1	100	0.86	3
1	100	0.5	3
3	10	1	3
3	10	2	3
3	50	1	3
3	25	1	3
3	25	0.79	3
3	25	0.5	3
3	25	0.5	9

Table 3.2   Data Sets for Feasibility Test

4   Results and Discussion

    Generally the results are given in tables indicating
    the test case, the gas outlet temperature, the solid
    face area and back side temperature and the effi-
    ciencies for evaluation. For all analysed cases the
    temperature distribution along the wall and the gas
    flow is plotted in the figures 1 to 20. Table 4.1
    gives the pretest results, table 4.2 the results of
    feasibility tests.

4.1   Evaluation Method of the Results

    For the evaluation of the results two efficiency
    factors were defined, their product giving the
    thermal efficiency, which is taken as evaluation
    criterion.

    The first one deals with the "reflection" efficiency,
    accounting for direct reflection losses on the front
    plate of the face area to the cold environment
    defined by

$$\eta_B = \frac{\text{incident solar field flux - reflected flux}}{\text{incident solar field flux}}$$

    the second one accounts for radiative losses of the
    face area and the orifice back to the black and
    cold environment, i. e. defined by

$$\eta_R = \frac{\text{incident solar flux - back radiated flux}}{\text{incident solar flux}}$$

The evaluation criterion for thermal efficiency is
then defined by

$$\eta_T \quad = \quad \eta_B \cdot \eta_R$$

The backscattering or reflection efficiency can be
calculated by

$$\eta_B \quad = \quad \frac{\dot{q} - \dot{q}_B}{\dot{q}}$$

$$\dot{q}_B \quad = \quad \dot{q} \, (1 - \varepsilon) \, \frac{A_F - A}{A_F}$$

$$\dot{q}_B = \dot{q} \, (1 - \varepsilon) \, \frac{(s + b)^2 - s^2}{(s + b)^2}$$

$$\eta_B \quad = \quad 1 - (1 - \varepsilon) \, \frac{(s + b)^2 - s^2}{(s + b)^2}$$

showing the dependence on the emissivity of the
ceramic and the geometry of the channel, i. e. the
wall thickness and the channel width. The following
two figures give the relationship "reflection"
efficiency with channel width, parameter: emissivity
and wall thickness:

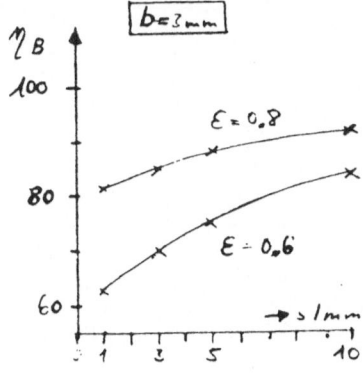 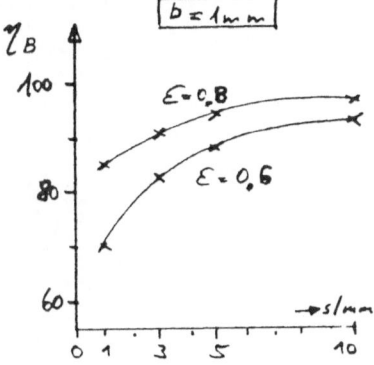

As a result: small cross-section length of the orifice and low emissivity of the ceramic lead to low "reflection" efficiencies.

The "radiation" efficiency η_R can be calculated by the ratio of air heating attained in each test case with respect to the maximum air heating achievable with given mass flow and at sufficient channel length, if all incident radiation would have been transformed into gas heating (theoretical limit).

$$\eta_R = \frac{\Delta T_A}{\Delta T_{A \text{ maximum (theoretical)}}}$$

In the parameter tests analysed the "radiation" efficiency is calculated by

$$\eta_R = \frac{T_{A \text{ out}} - T_{A \text{ in}}}{980 \text{ K} \cdot \dot{m}_{St}/\dot{m}_{A_F}}$$

The efficiency factors, thus defined, were put into the tables of the results of the various test cases.

| Case | Fig. | Channel width | Air Outlet $T_{A\,out}$ | Ceramic $T_{s\,in}$ | $T_{s\,out}$ | η_R | η_B | η_T |
		s/mm	°C	°C	°C	%	%	%
Pretest 1	1	1	709	1332	1268	70	81	57
"Geometry"	3	3	732	1294	1088	73	85	62
λ_s = 50 W/mK	4	5	680	1370	1017	67	88	59
ε^s = 0.6	5	10	556	1519	934	55	92	50
α-formulas								
Cosine								
distribu-								
tion								
Pretest 2								
"Phys.	7	50 % λ_s	692	1358	994	69	85	58
Properties"	6	200 % λ_s	758	1252	1134	75	85	64
s = 3 mm	8	ε = 0.6	762	1334	1113	76	70	53
Cosine	11	50 % α	440	1609	1479	43	85	37
distribu-	12	200 % α	854	1067	926	85	85	72
tion								
Pretest 3		Parallel Radiation						
"Radiation								
distribu-								
tion"								
s = 3 mm	9	2.9°	712	1227	1076	(71)	85	60
λ_s = 50 W/mK	10	5.7°	731	1250	1098	(73)	85	62
ε^s = 0.8	13	30°	730	1292	1077	72	85	62
α-formulas								

Pretest Results   Table 4.1

Results of Pretests
         (Table 4.1)

         Pretest 1

         Generally, the channel length was not long enough for
         the test cases, because the gas outlet temperature
         did not reach the back side wall temperature. The
         channels with smaller channel width show the relative
         highest efficiency. For further analysis the orifice
         cross-section length of 3 mm was chosen as "standard"
         case because of its highest efficiency. The temperature
         distribution along the channel is shown in Fig. 3.

         Fig. 21 shows the change of air velocity, Reynolds-number
         and of convective heat transfer coefficients along the
         channel for the "standard" case as an example of simi-
         lar distributions for all test cases. The gas velocity
         rises with temperature rise in the channel, because
         of gas expansion; the Reynolds-number becomes lower
         because of higher air-viscosity with higher tempera-
         ture and the convective heat transfer coefficient
         follows mainly the entrance region law and the
         Reynolds-number decrease, but is influenced too by
         higher thermal conductivity with temperature rise and
         less influence of the temperature correction factor
         with approaching wall and gas temperatures. There-
         fore the α-coefficient shows a minimum value. The face
         area convective heat transfer coefficient is consider-
         ably lower than that of the inlet flow; hence the
         two different correlations used (Chapter 2.2.2) seem
         to be somewhat mismatched.

Higher ceramic conductivity is favourable with respect
to better efficiency; thermal conductivity as low as
25 W/mK, i. e. Si_3N_4 is not suitable.

The emissivity reduction leads to less radiation to the
environment, therefore the "radiation" efficiency is
higher but the face temperature is higher. The reflection losses at the face area are considerably
higher, giving lower overall efficiency.

The variation of the convective heat transfer coefficients shows the importance of these values for the result.
With higher values, the ceramic temperature is considerably lower and the approach of the gas outlet temperature
is better. The inverse is true for lower α's.

Because of rather high resulting face temperatures and
the strong influence of the convective heat transfer,
the used α-correlations should be carefully examined.
There are very steep temperature gradients of the
inlet flow, and very sharp α- and Re-changes at short
distances, and the correction factor for physical
properties is out of range (Chapter 2.2.2) at the
channel inlet.

Therefore it is necessary to verify the convective
heat transfer at realistic channel conditions for the
inlet flow situation theoretically with thermohydraulic
codes and/or by experiments.

Pretest 3

Regarding the thermal efficiencies there seem to be
not much difference between the results of variations
on incident radiation, either

-   cosine distribution or

-   parallel rays, inclined to the channel axis by
    different angles.

But for very small inclination angles it must be re-
minded that part of incident radiation is lost for thermal exchange with the gas at the back side of the
channel; the efficiency is expected to be greater
at longer channels. For the 30 degree inclination of
parallel rays the results are nearly identical with
the cosine distributed radiation.

It can be concluded that the cosine distributed radiation is the appropriate one to analyse the receiver
feasibility in these tests. Nevertheless, it is probable
that for parallel radiation higher thermal efficiencies
are achievable.

Case Fig.	Channel width s/mm	Ratio length L to s L/s	Relative mass flow $\dot{m}_{AF}/\dot{m}_{St}$	Air outlet $T_{A\,out}$ °C	Ceramic face $T_{S\,in}$ °C	Ceramic back $T_{S\,out}$ °C	η_R %	η_B %	η_T %
Wall thickness b = 3 mm									
1	1	10	1	709	1332	1268	70	81	57
2	1	100	1	857	1059	858	85	81	69
20	1	100	0.86	956	1124	956	82	81	66
15	1	100	0.5	1284	1360	1284	64	81	52
3	3	10	1	732	1294	1088	73	85	62
14	3	10	2	411	1225	1006	80	85	68
–	3	50	1	794	1192	796	79	85	67
18	3	25	1	790	1200	831	79	85	67
19	3	25	0.79	936	1263	964	74	85	63
16	3	25	0.5	1219	1402	1222	61	85	52
Wall thickness b = 9 mm									
17	3	25	0.5	994	1541	1237	50	81	40

Feasibility Tests

Table 4.2

4.3 Results of Feasibility Tests
(Table 4.2)

For the 1 mm channel a length of 100 mm was chosen instead of 10 mm and the maximum thermal efficiency could be achieved. The gas outlet temperature has already reached the wall temperature at about 50 mm channel length (Fig. 2); therefore a shorter channel length is sufficient (not important for the thermal efficiency but for pressure loss, see below). Reduction of mass flow rate leads to higher gas outlet temperatures, but to higher ceramic temperatures and lower thermal efficiency.

The results for the 3 mm channel are very similar, but with higher ceramic temperatures and less efficiency. The results are plotted in Fig. 22 to be able to interpolate for 1000 °C gas outlet temperature:

Channel width	L/s	Temperatures Gas outlet	Ceramics front (maximum)	Efficiency	Mass flow rate
		°C	°C	%	kg/s m²
s = 1 mm	100	1000	1150	64	0.75
s = 3 mm	25	1000	1290	61	0.67

Concerning the ratio of channel length to width at s = 3 mm, it would be sufficient to have channel lengths between 75 mm (L/s = 25) and 150 mm (L/s = 50), because the gain in efficiency for longer channels is not important (at least with cosine distributed radiation).

There was one test at L/s = 10 for s = 3 mm channel width with two times the standard mass flow, which means that the gas temperature rise has to be compared with 490 K, the maximum theoretical value achievable without radiative losses.

It is obvious that gas outlet and channel back end temperature are still mismatched at this channel length, therefore the efficiency is underestimated in this case. By direct comparison with the standard case it is shown that much lower gas temperatures do not lead to a similar reduction in the ceramic temperatures.

Concerning the wall thickness, it was found that longer channel length is necessary to reach the maximum gas outlet temperature (comparison for s = 3 mm, L/s = 25, 50 % standard mass flow rate between the case with 3 or 9 mm wall thickness) and that the face ceramic temperature and the temperature difference along the channel wall will be higher.

Pressure Loss

As indicated in the formula used for pressure loss calculation (Chapter 2.2.3), the main pressure loss is to be expected by friction losses, which in turn are dependent on the channel geometry, i. e. the ratio channel length to hydraulic diameter L/s, and the Reynolds-number distribution along the channel, i. e. the flow rate, respectively the velocity field and the temperature dependent gas viscosity.

In the preceding parameter tests always the same
mass flow rate $\dot{m}_{St}$ = 0.934 kg/m^{2}s was used as standard. The corresponding flow rate in the channels is
given in Table 4.3.

Channel	Flow rate 10^{-5} kg/s	Case	Pressure loss Pa
s = 1 mm	1.5	L/s = 100	6080
s = 3 mm	3.36	L/s = 10	31
		L/s = 25	102
s = 5 mm	5.98	-	-
s = 10 mm	15.8	-	-

Table 4.3   Mass Flow and Pressure Loss in the
             Channels

The pressure loss, calculated for some cases, are
indicated in this table, too.

The mean gas velocity in the channels are about
30, 8, 5 or 3 m/s corresponding to 1, 3, 5 or 10 mm
channel width (L/s = 10).

The mass flow rate was varied only in the feasibility
tests. For two "suitable" channel geometries, the
pressure loss was drawn with relative mass flow rate in
Fig. 23. For the channel with s = 1 mm, L/s = 100 it is
possible to reduce the pressure loss approximately to one
half of the indicated value, if the channel length would
be reduced to one half, too, which would be sufficient
for complete heat exchange (Fig. 15, 20).

For further evaluation of the efficiency of volumetric receivers made of ceramics it must be kept in mind that the calculated parameter tests were made

- with an idealised section of the receiver, i. e. no boundary losses at the outer surfaces, except the face area, were taken into account;

- with an idealised solar flux field fixed at 100 W/cm^2, i. e. incident solar flux was taken as overall constant, without attenuation at its borders;

- with a limited parameter set, to fulfill the goal of "feasibility" and to allow a relative evaluation;

and therefore the results must be taken as a first approach to find appropriate receiver geometries under the conditions taken into account.

To find out the appropriate receiver geometry for a prototype ceramic receiver to be tested in a real solar test field it is recommended to do in a first step a complete sensitivity study, with systematic variation of all important parameter sets with the same idealised model to find out more precisely the most suitable one. For the parameters channel width, wall thickness and thermal conductivity precise data have to be given by the manufacturer.

In a second step, a complete receiver model for thermal analyses could be built up to take into account border effects, the real solar field and appropriate gas flow for steady-state and transient conditions.

Because of the important influence of the convective heat transfer on the ceramic temperatures and due to the lack of reliable heat transfer correlations for the face area and the entrance region of the channel it is necessary to analyse this situation by thermo-hydraulic methods before experimental tests.

It should be kept in mind that besides the thermo-hydraulic effects

-    the thermo-mechanical properties and

-    the pressure loss of the air flow with respect
     to an appropriate heat exchanger circuit

have to be evaluated.

6 <u>Conclusion</u>

It has been shown with numerical treatment of thermal
interchange between radiative, convective and con-
ductive heat transfer that for incident solar flux
of 100 W/cm^2 a volumetric receiver made of SiSiC
ceramics with thermal conductivity of 50 W/mK,
emissivity of $\varepsilon = 0.8$, wall thickness 3 mm, in an
open air flow of 1000 °C outlet temperature
is feasible with the following characteristics:

Channel geometry Width s	Ratio length to width L/s	Maximum ceramic tempera- ture °C	Mass flow rate kg/m^{2}s	Pressure loss Pa	Efficiency %
1	100 *)	1150	0.75	6080 *)	66
3	25	1290	0.67	31	63

Because of the relatively high face temperatures of
the ceramic near the melting point of Silicon in SiSiC
and the strong influence of the convective heat transfer
coefficient on it, it is necessary to verify the used
α-correlation theoretically with thermohydraulic codes
and/or by experiments.

The proposed arrangement is probably not a finally optimised one, but indicates the feasibility for the investigated conditions.

---

*) L/s 50 should be sufficient for thermal exchange with an
approximate pressure loss reduction to about 30 mbar.

7 References

/ 1 /   Grönefeld, G.
"Programm LIWAK"
Interatom-Notiz Nr. 35.02353.3 dated 25.11.85

/ 2 /   Müller, K.
"ALBEMO, ein Programm zur Berechnung des
Strahlungstransportes in Hohlräumen für
Neutronen-, Gamma- und Wärmestrahlung"
Interatom-Bericht Nr. 70.03565.6 dated 12.10.85

/ 3 /   Churchill, S. W. and Bernstein, M.
A Correlating Equation for Forced Convection
from Gases and Liquids to a Circular Cylinder
in Crossflow,
J. Heat Transfer, Vol. 99, pp. 300 - 306, 1977
cited in J. P. Holman, Heat Transfer,
5th Edition, 1981, McGraw-Hill

/ 4 /   Kays, W. M.
Trans. ASME, 77, 1265 (1955),
cited in / 7 /

/ 5 /   VDI-Wärmeatlas, 2. Auflage, 1974
p. Gb 5

/ 6 /   Idelchik, I. E.
Handbook of Hydraulic Resistance,
2. Edition, 1986, Hemisphere Publishing
Corporation

/ 7 /   Rohsenow, W. M. and Choi, H. Y.
Heat, Mass and Momentum Transfer, 1961
Prentice-Hall Inc.

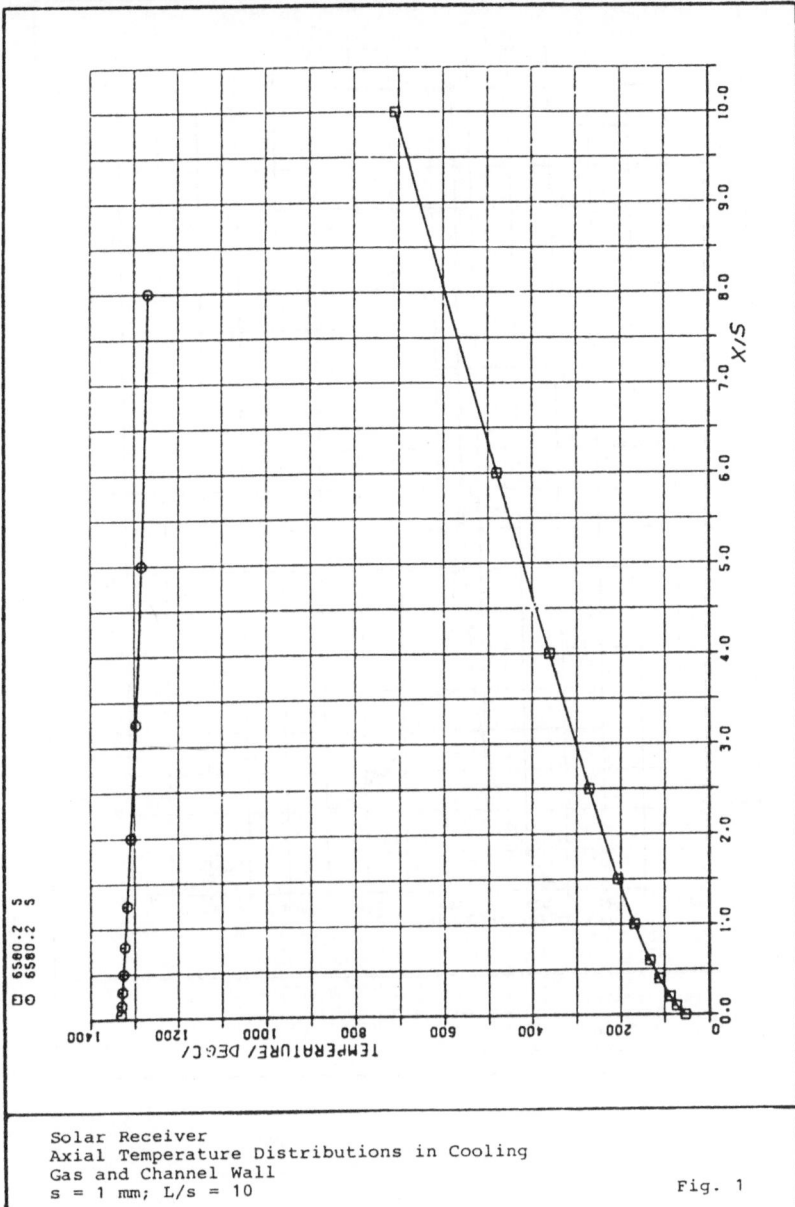

Solar Receiver
Axial Temperature Distributions in Cooling
Gas and Channel Wall
s = 1 mm; L/s = 10

Fig. 1

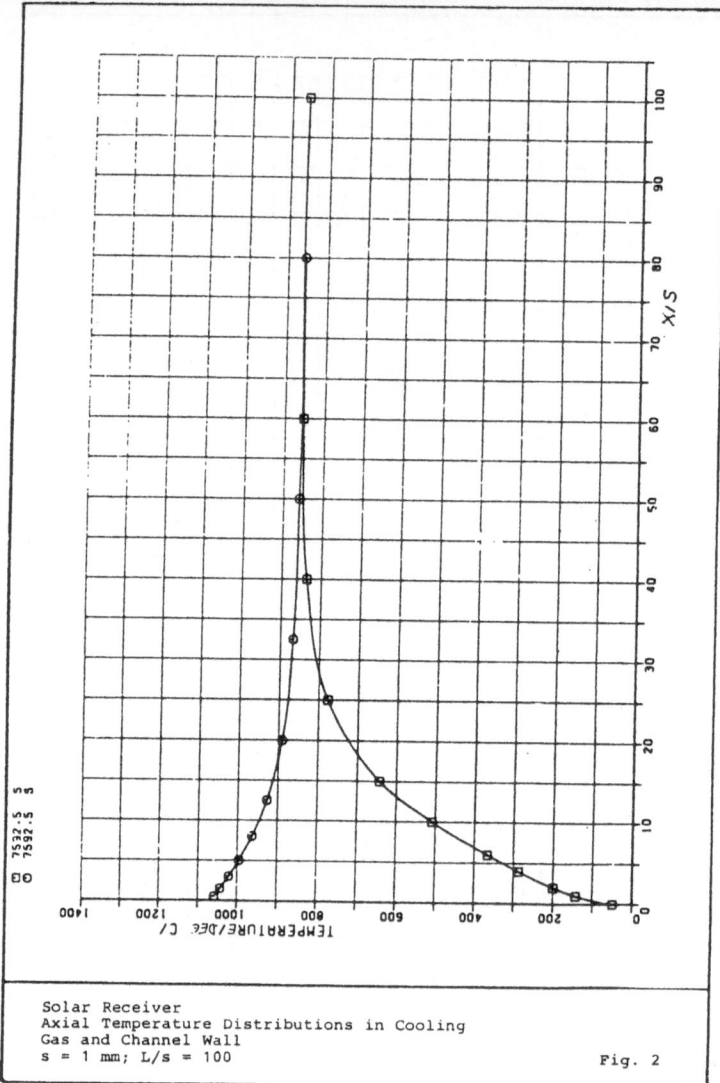

Solar Receiver
Axial Temperature Distributions in Cooling
Gas and Channel Wall
s = 1 mm; L/s = 100

Fig. 2

- 34 -

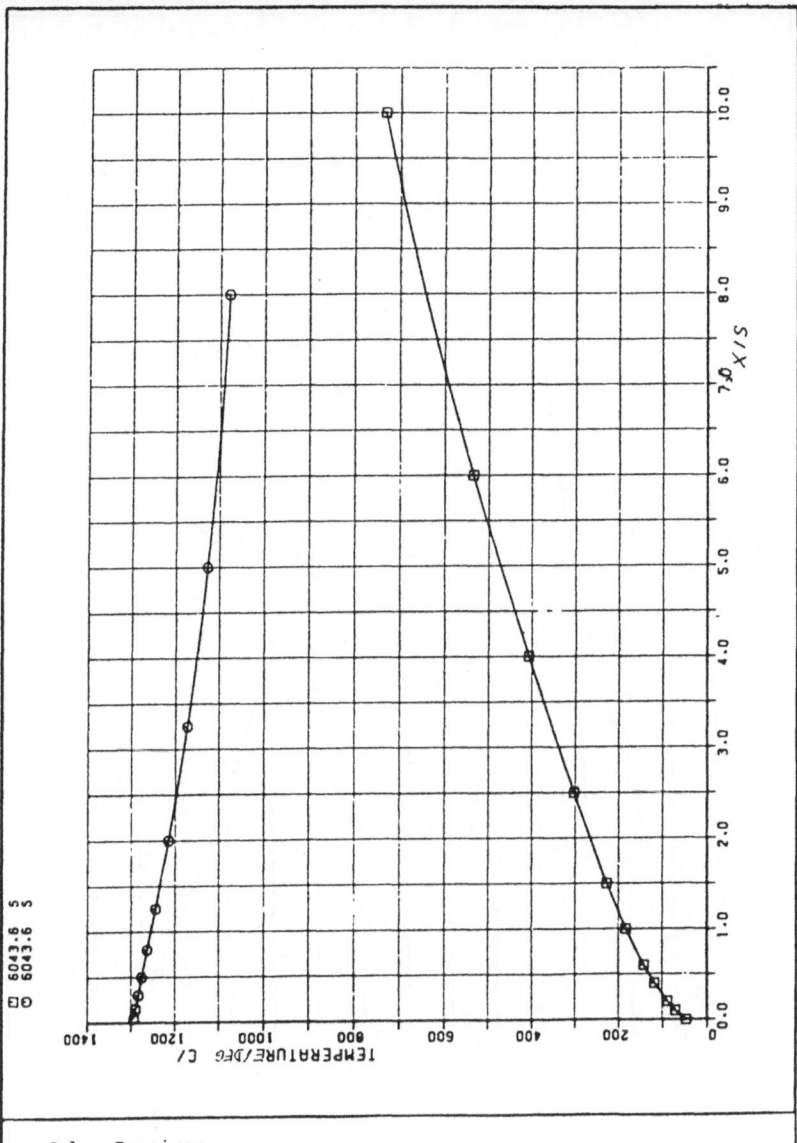

Solar Receiver
Axial Temperature Distributions in Cooling
Gas and Channel Wall
s = 3 mm; L/s = 10

Fig. 3

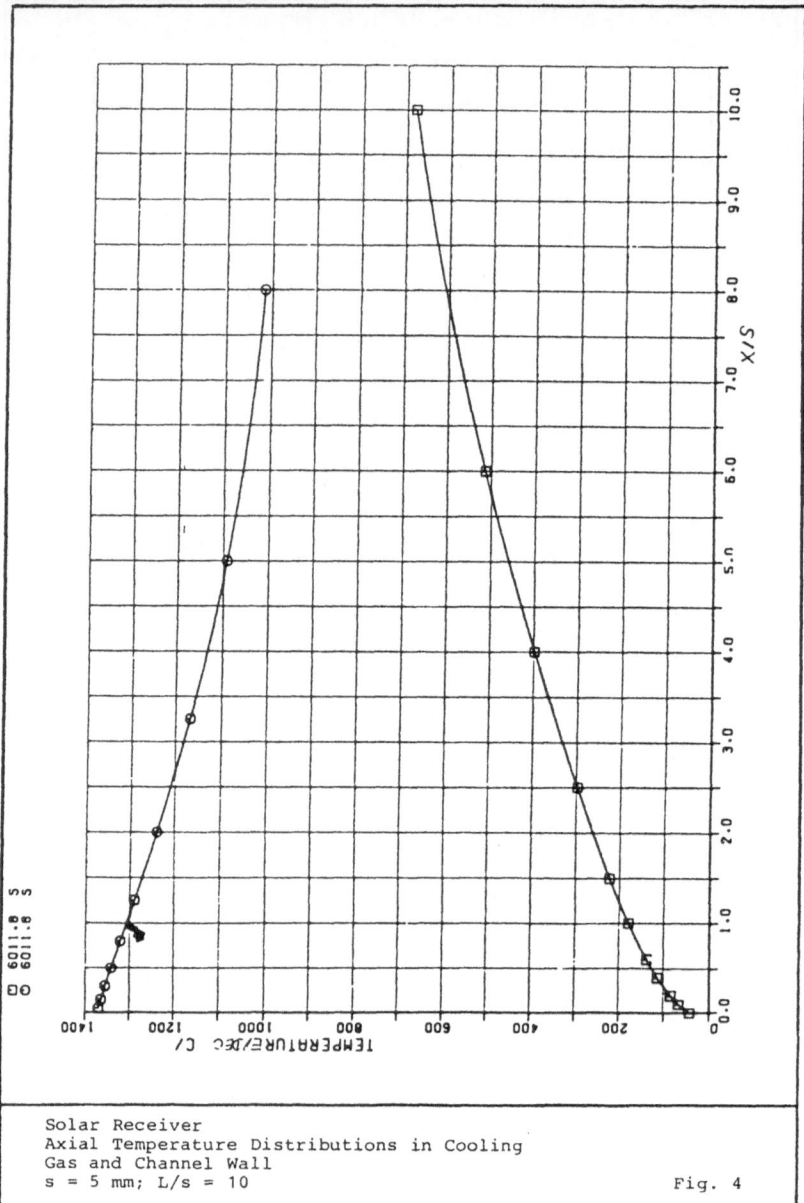

Solar Receiver
Axial Temperature Distributions in Cooling
Gas and Channel Wall
s = 5 mm; L/s = 10   Fig. 4

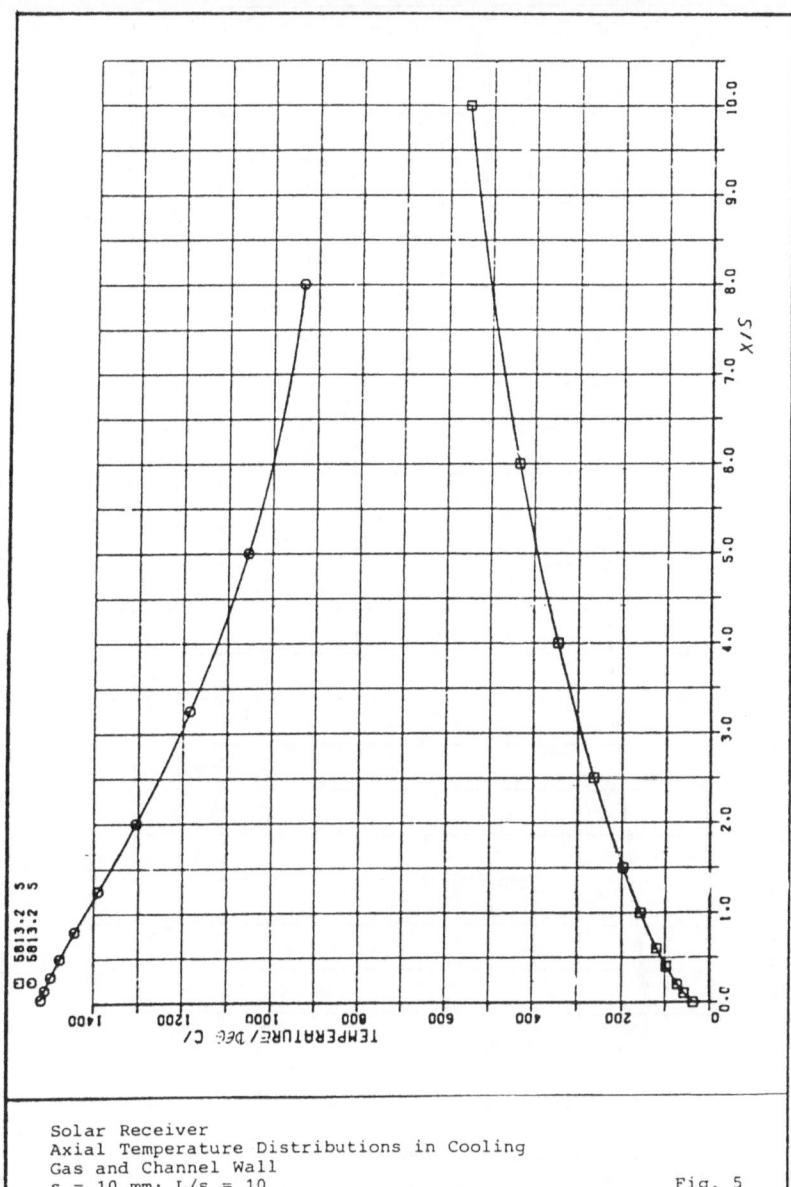

Solar Receiver
Axial Temperature Distributions in Cooling
Gas and Channel Wall
s = 10 mm; L/s = 10

Fig. 5

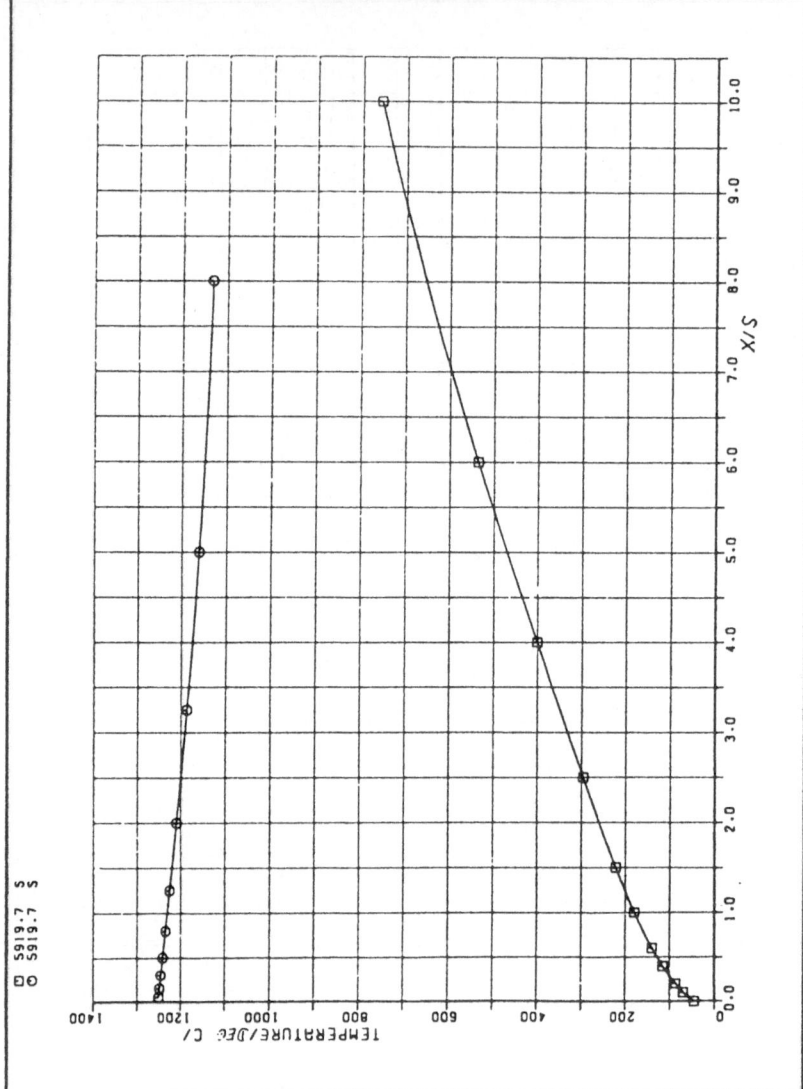

Solar Receiver
Axial Temperature Distributions in Cooling
Gas and Channel Wall
s = 3 mm; L/s = 10; λ = 100 W/(m·K)

Fig. 6

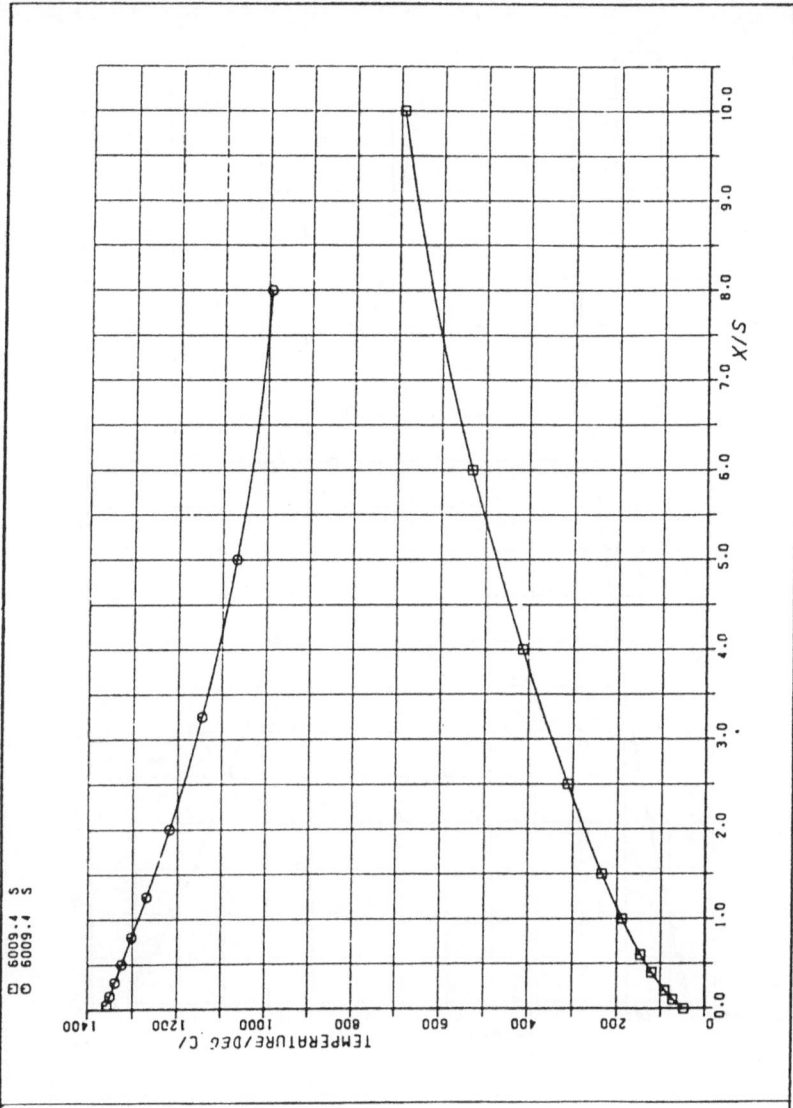

Solar Receiver
Axial Temperature Distributions in Cooling
Gas and Channel Wall
$s = 3$ mm; $L/s = 10$; $\lambda = 25$ W/(m·K)

Fig. 7

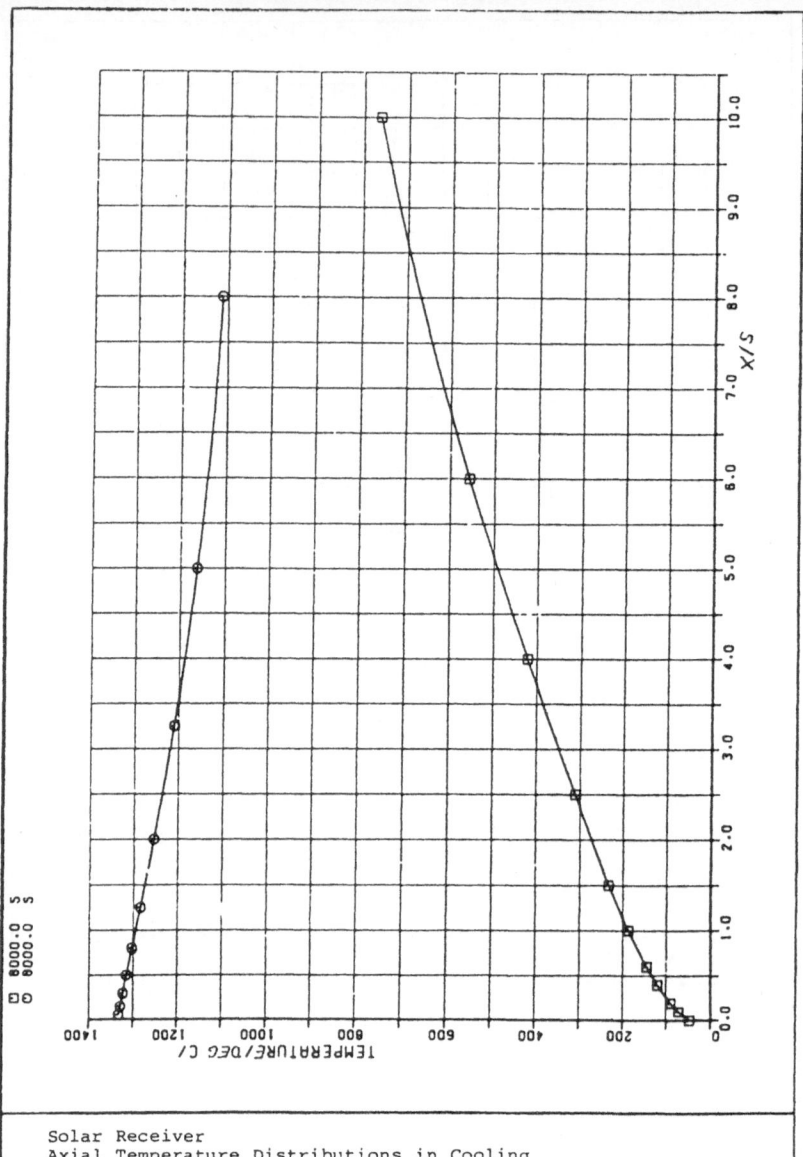

Solar Receiver
Axial Temperature Distributions in Cooling
Gas and Channel Wall
s = 3 mm; L/s = 10; ε = .6   Fig. 8

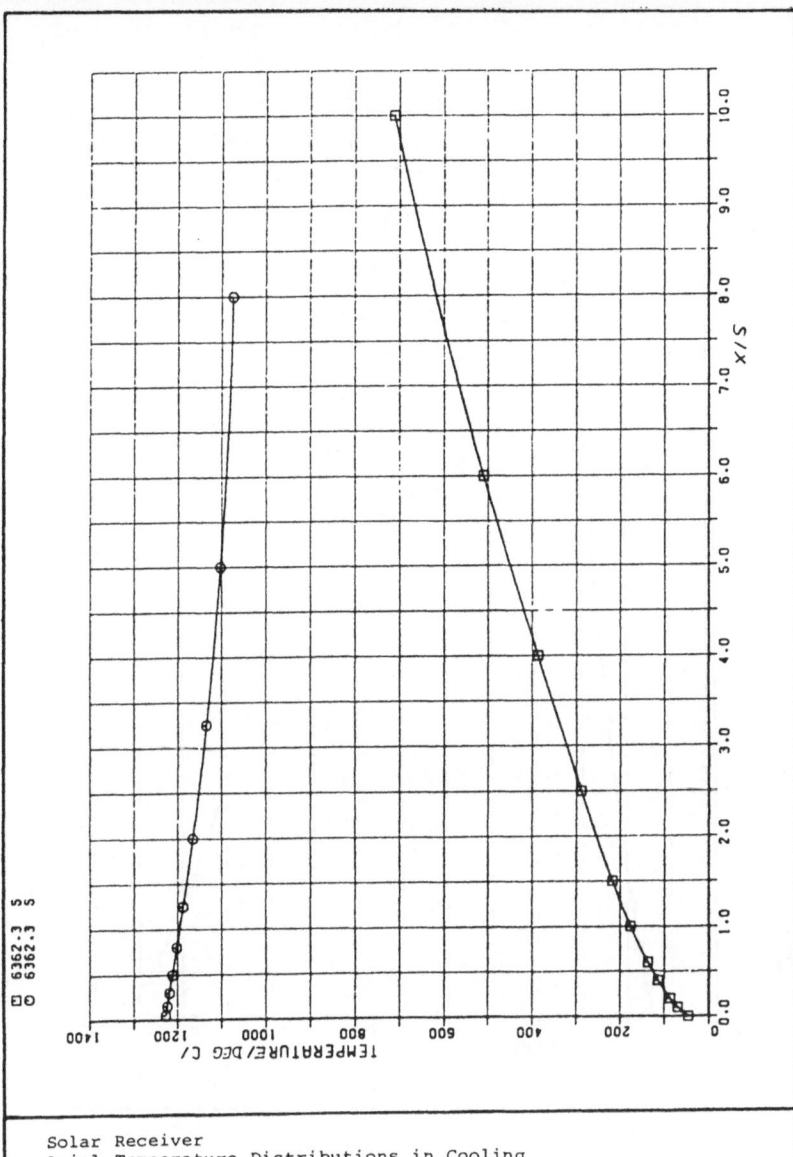

Solar Receiver
Axial Temperature Distributions in Cooling
Gas and Channel Wall
s = 3 mm; L/s = 10; Θ = 2.86°

Fig. 9

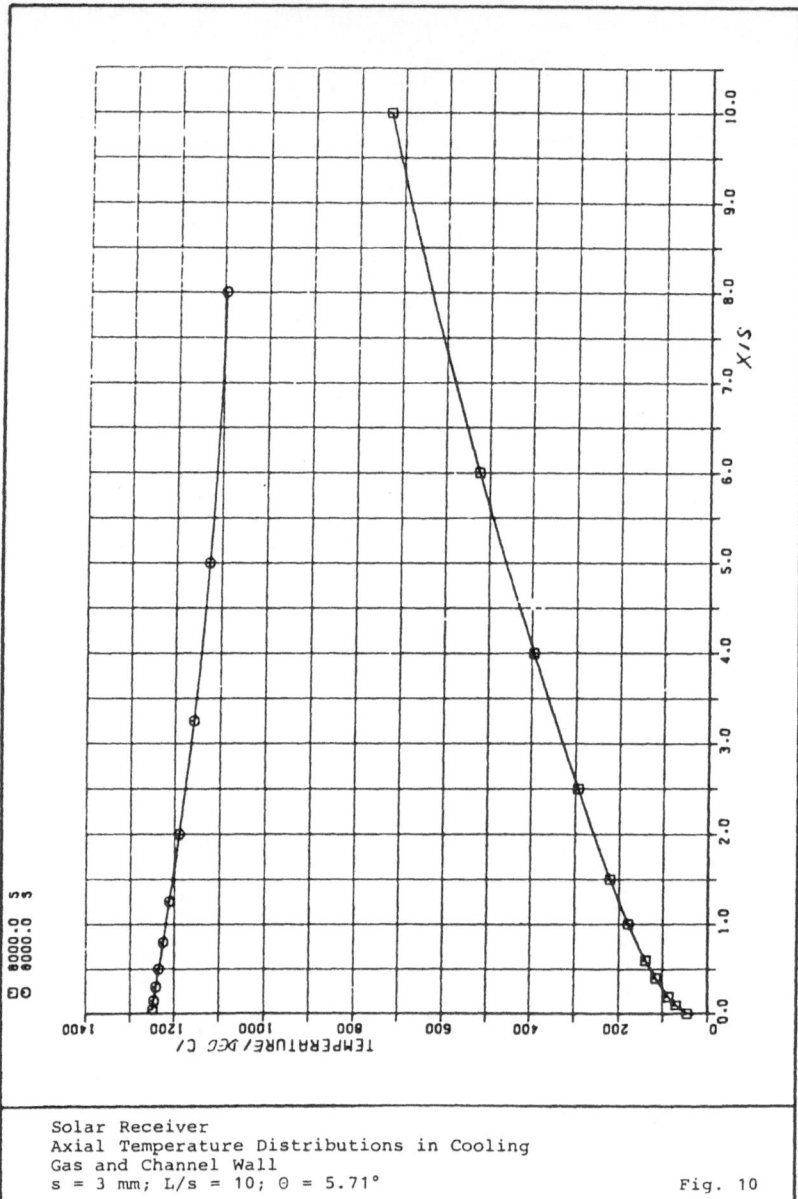

Solar Receiver
Axial Temperature Distributions in Cooling
Gas and Channel Wall
s = 3 mm; L/s = 10; Θ = 5.71°

Fig. 10

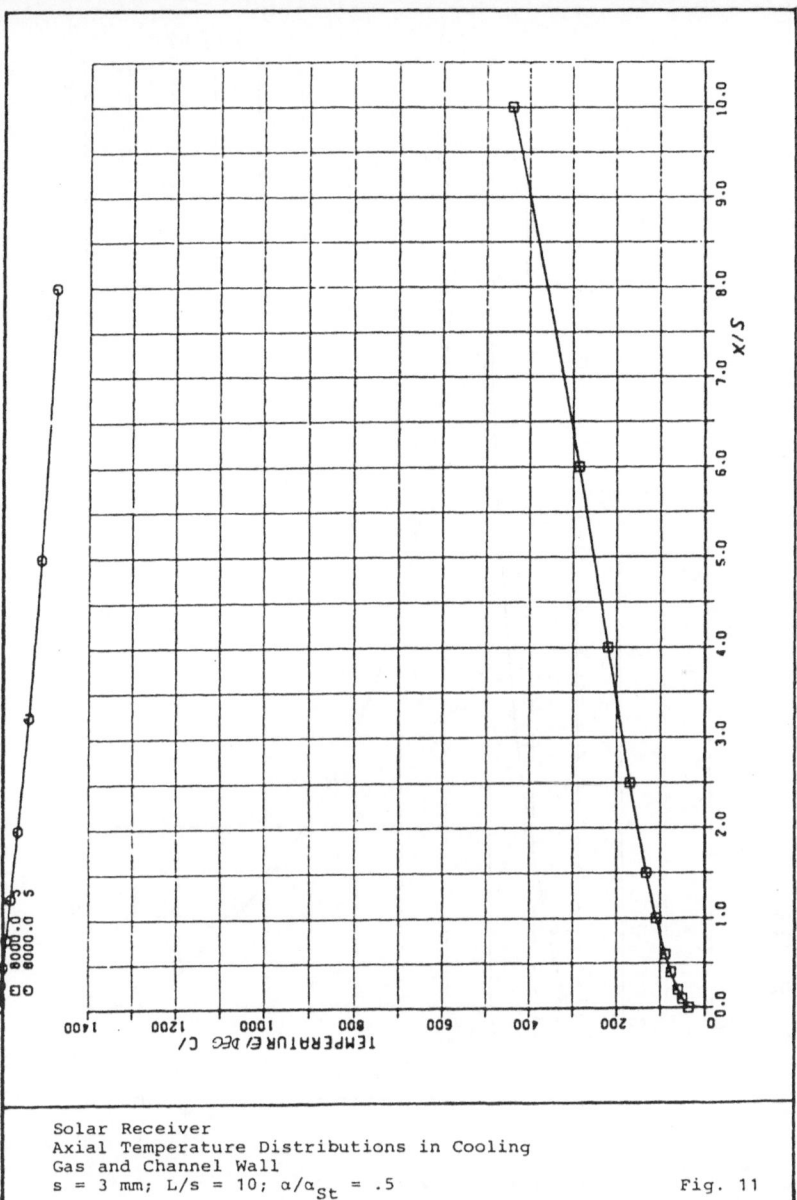

Solar Receiver
Axial Temperature Distributions in Cooling
Gas and Channel Wall
s = 3 mm; L/s = 10; α/α_{St} = .5   Fig. 11

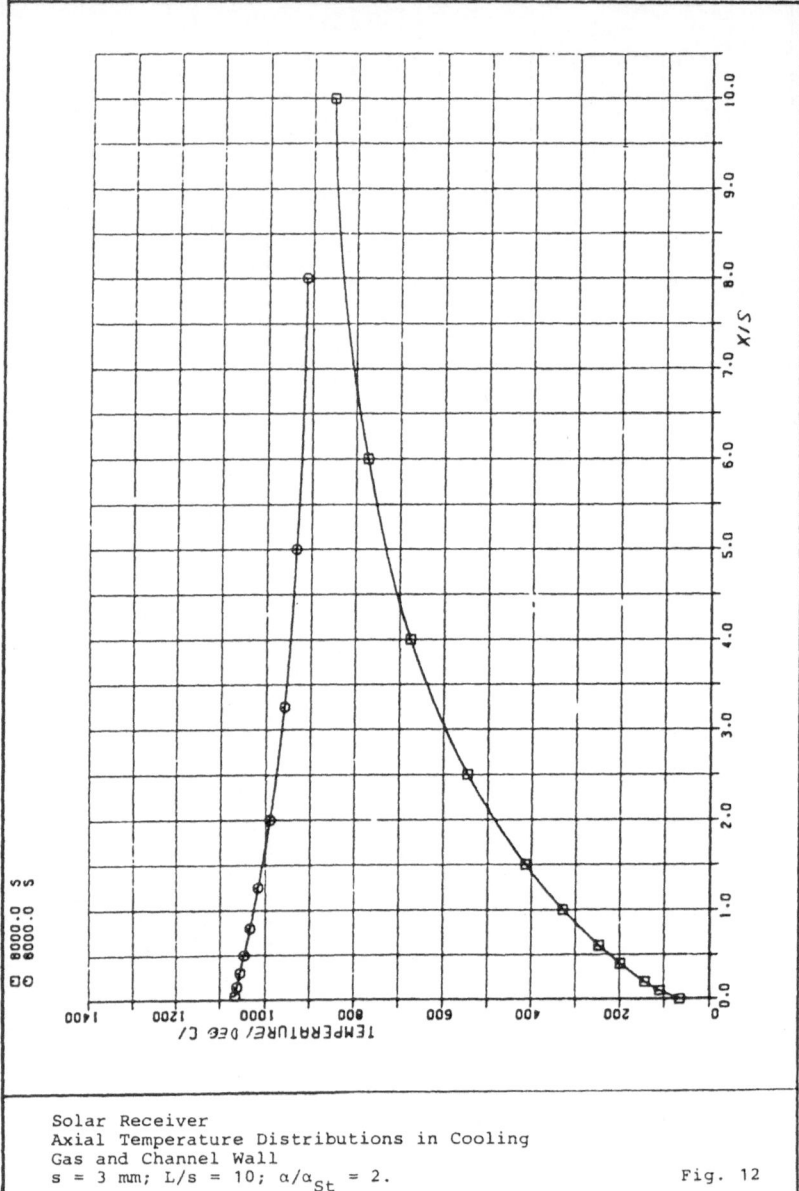

Solar Receiver
Axial Temperature Distributions in Cooling
Gas and Channel Wall
s = 3 mm; L/s = 10; α/α_{St} = 2.

Fig. 12

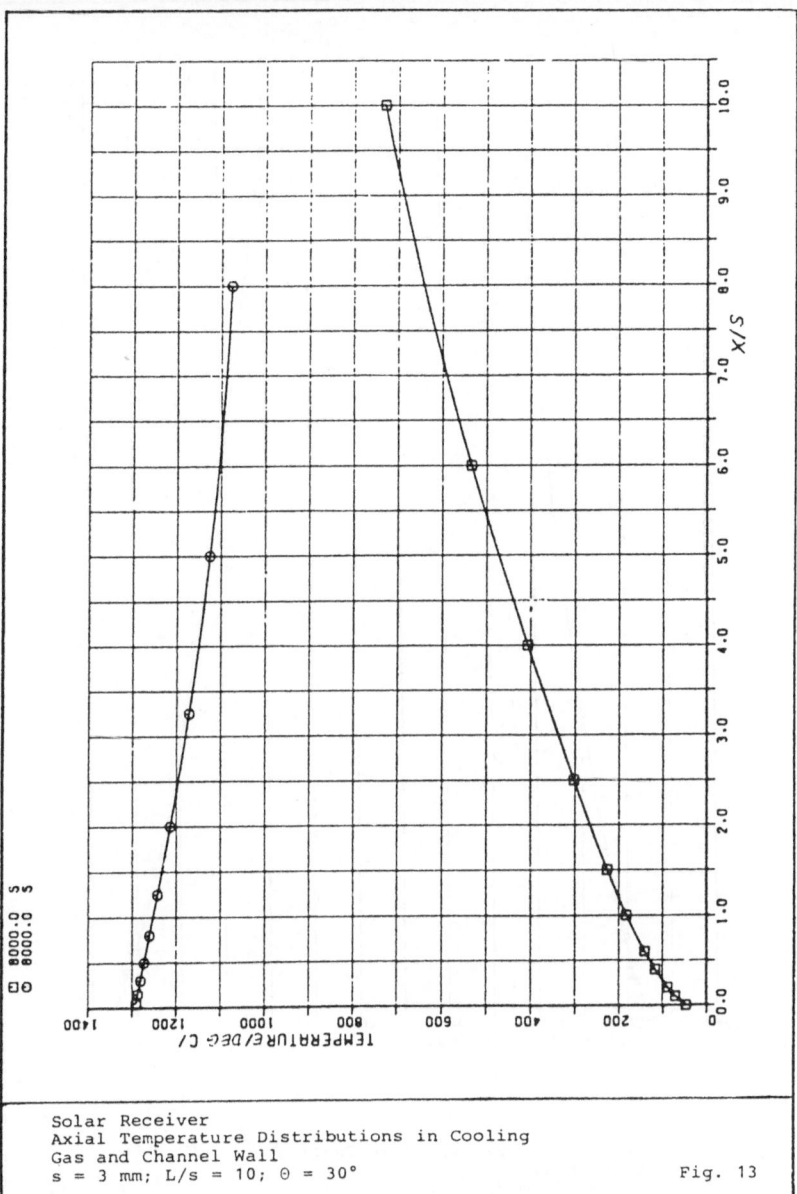

Solar Receiver
Axial Temperature Distributions in Cooling
Gas and Channel Wall
s = 3 mm; L/s = 10; Θ = 30°

Fig. 13

- 45 -

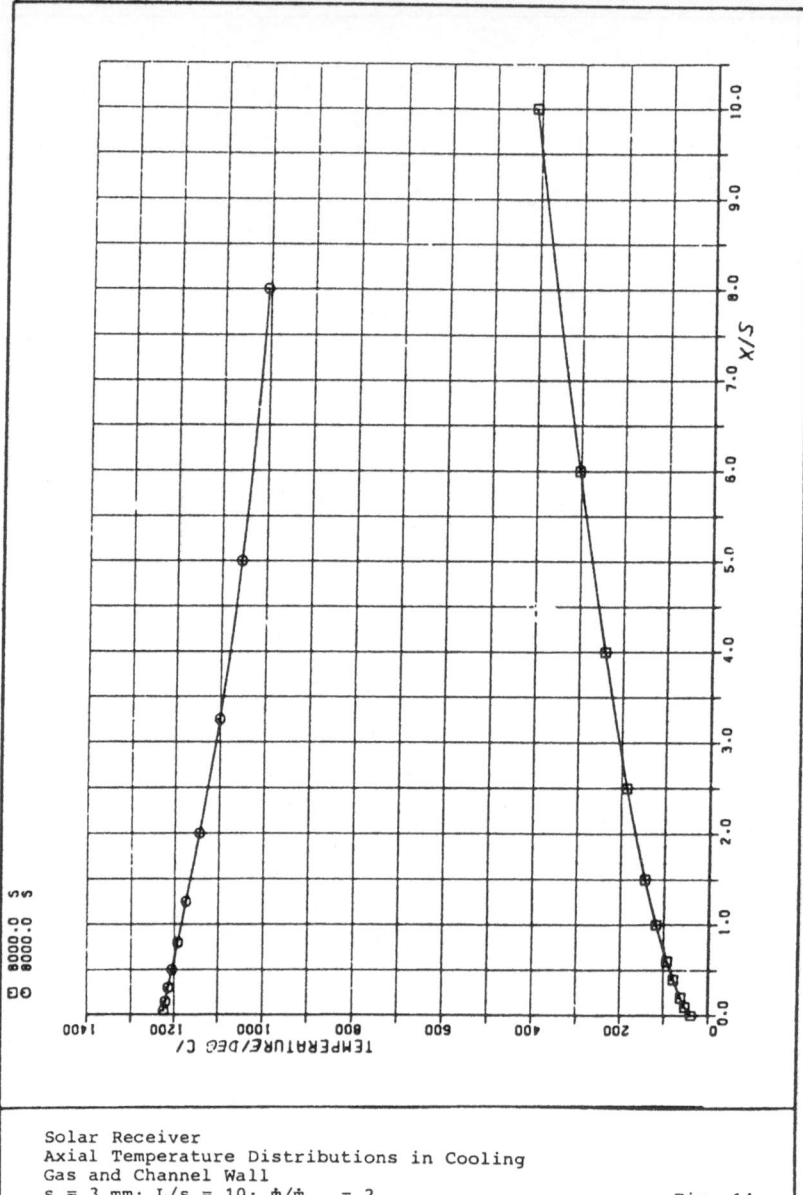

Solar Receiver
Axial Temperature Distributions in Cooling
Gas and Channel Wall
s = 3 mm; L/s = 10; $\dot{m}/\dot{m}_{St}$ = 2.

Fig. 14

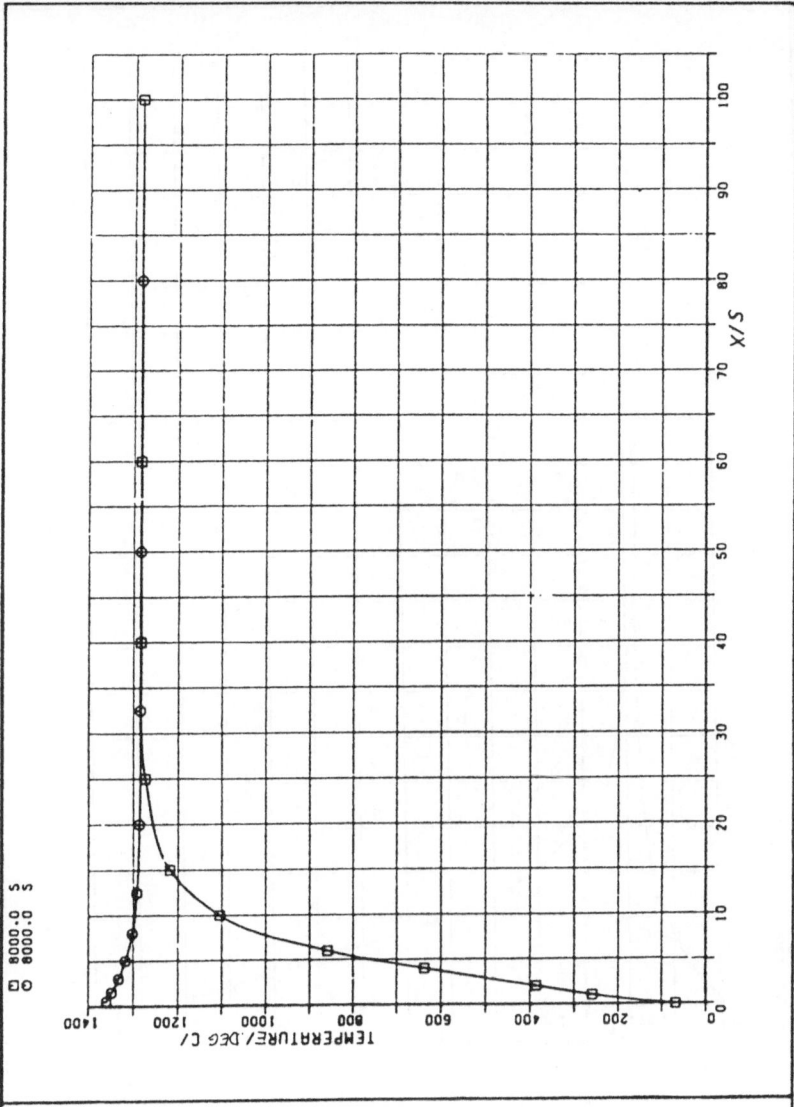

Solar Receiver
Axial Temperature Distributions in Cooling
Gas and Channel Wall
s = 1 mm; L/s = 100; $\dot{m}/\dot{m}_{St}$ = .5   Fig. 15

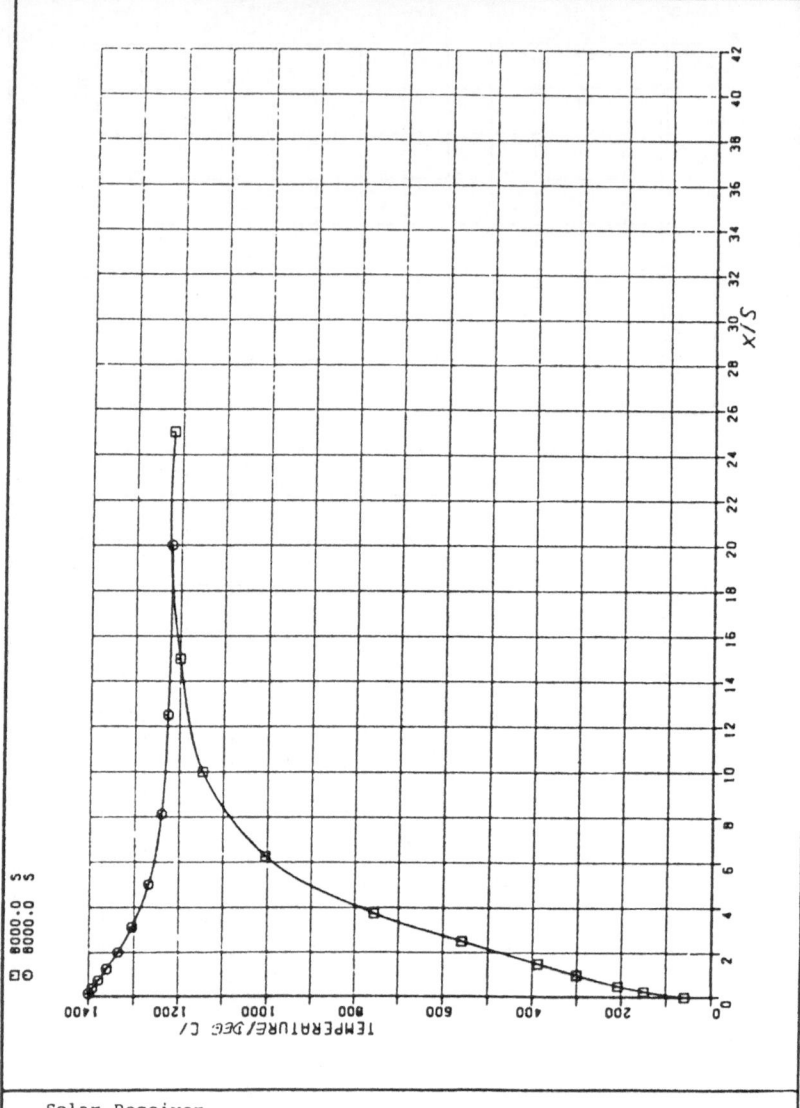

Solar Receiver
Axial Temperature Distributions in Cooling
Gas and Channel Wall
s = 3 mm; L/s = 25; $\dot{m}/\dot{m}_{St}$ = .5   Fig. 16

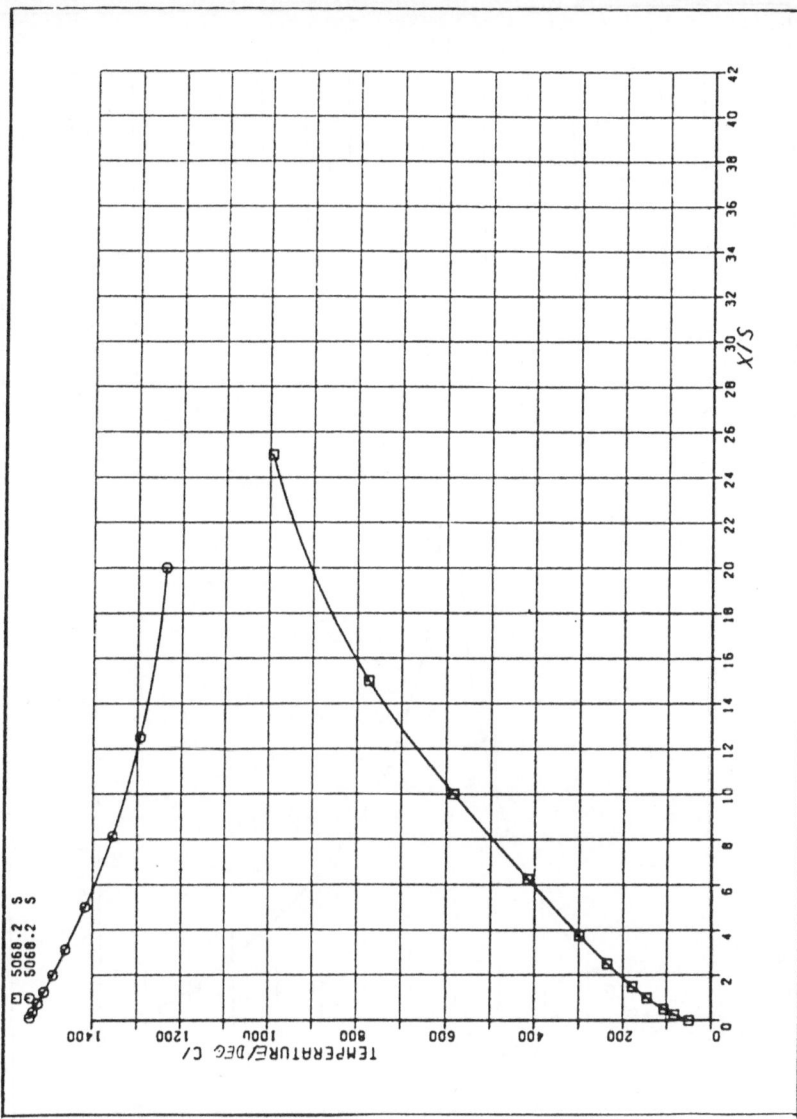

Solar Receiver
Axial Temperature Distributions in Cooling
Gas and Channel Wall, B/2 = 4.5 mm
s = 3 mm; L/s = 25; $\dot{m}/\dot{m}_{St}$ = .5

Fig. 17

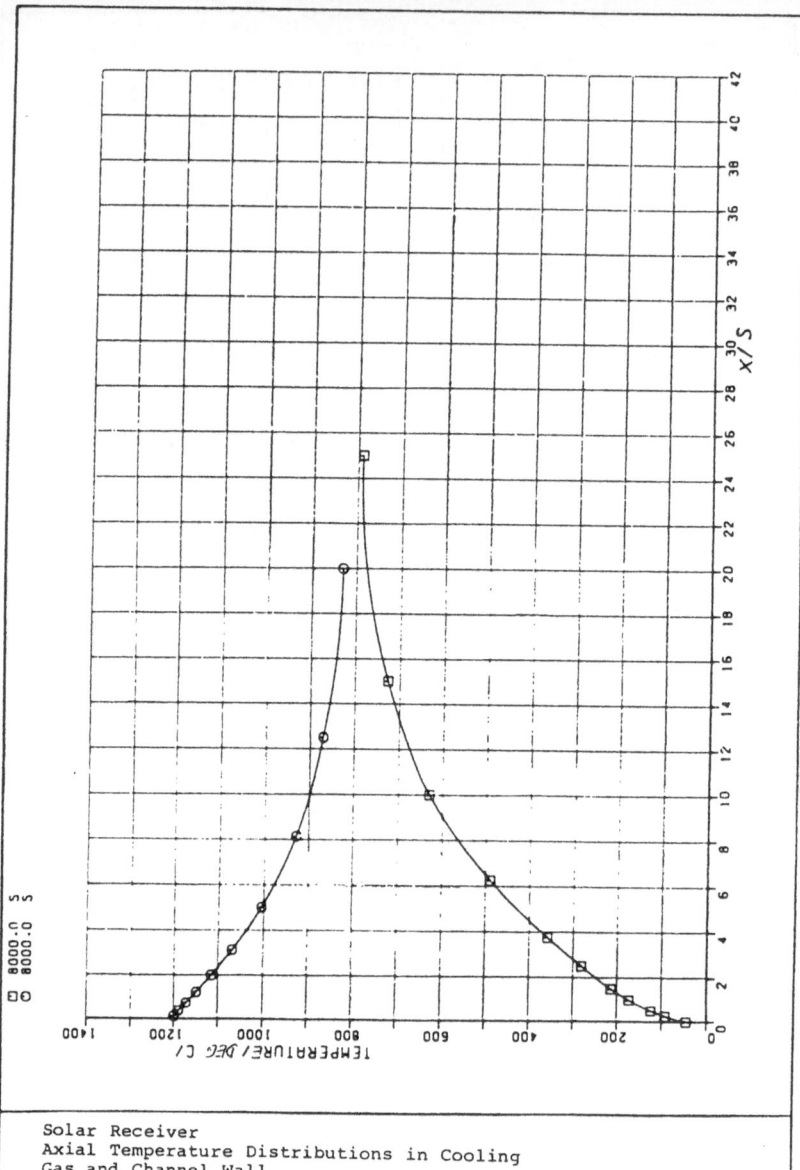

Solar Receiver
Axial Temperature Distributions in Cooling
Gas and Channel Wall
s = 3 mm; L/s = 25

Fig. 18

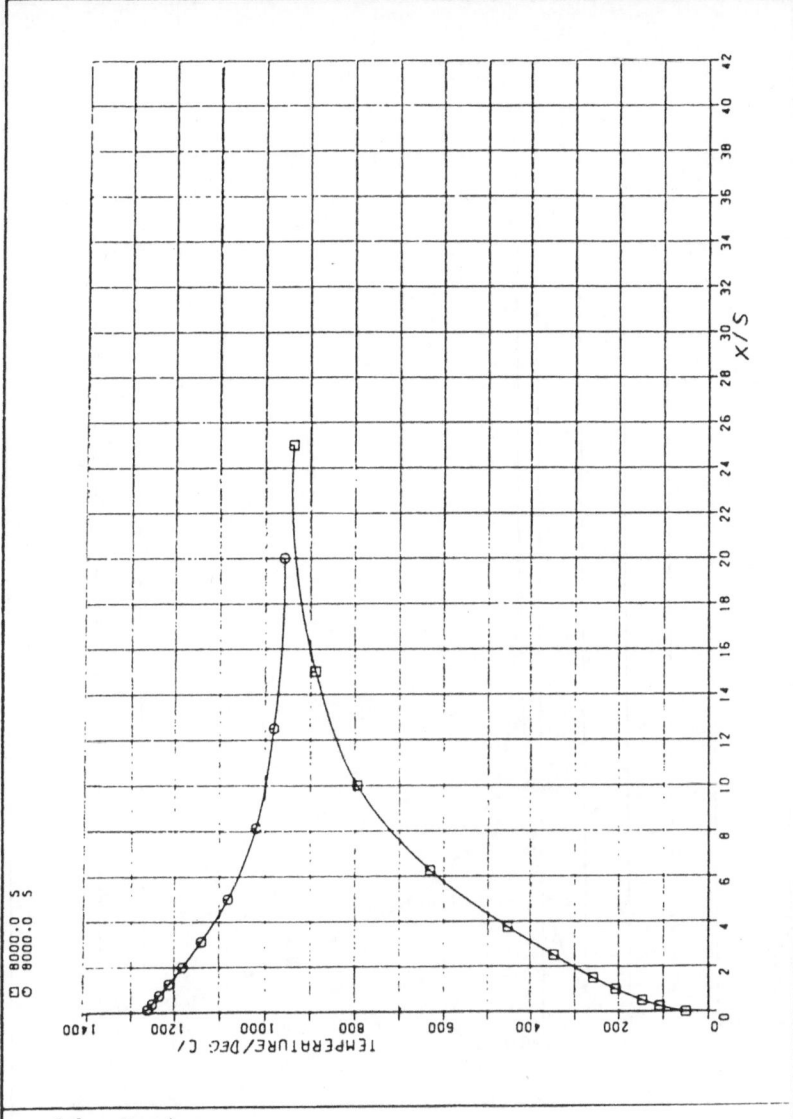

Solar Receiver
Axial Temperature Distributions in Cooling
Gas and Channel Wall
s = 3 mm; L/s = 25; $\dot{m}/\dot{m}_{St}$ = .79

Fig. 19

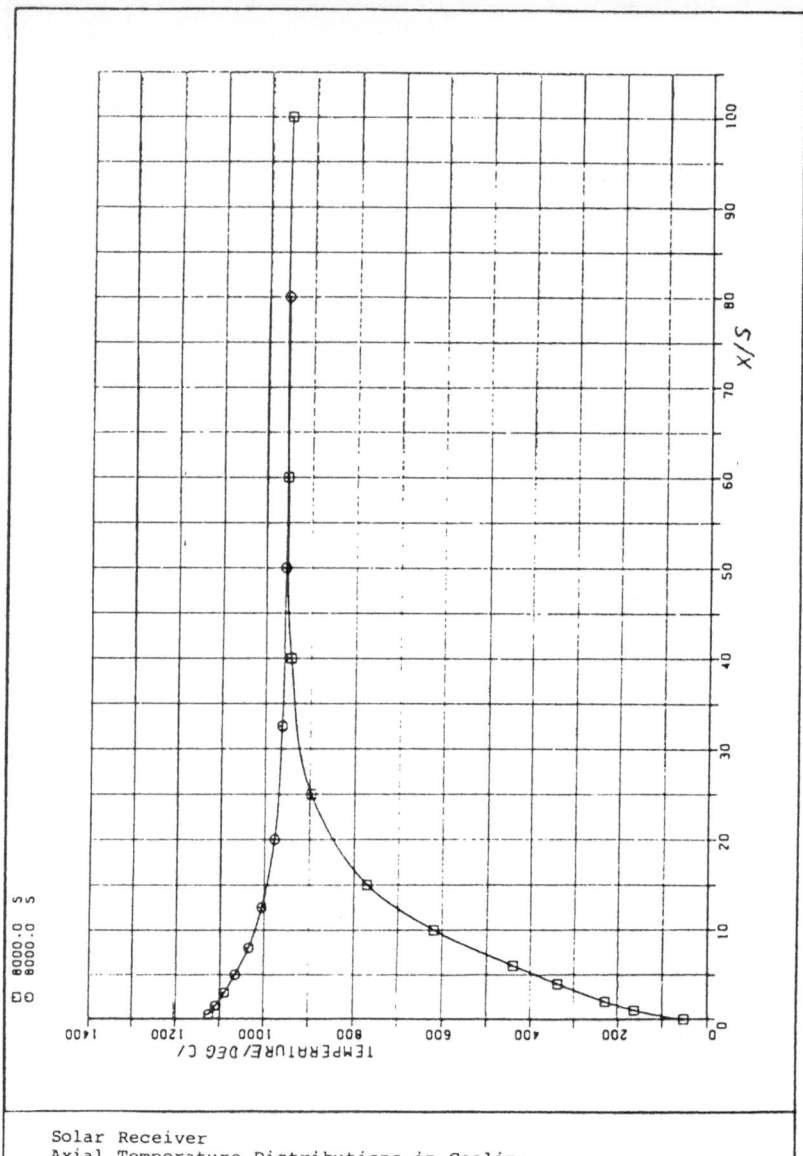

Solar Receiver
Axial Temperature Distributions in Cooling
Gas and Channel Wall
s = 1 mm; L/s = 100; $\dot{m}/\dot{m}_{St}$ = .86

Fig. 20

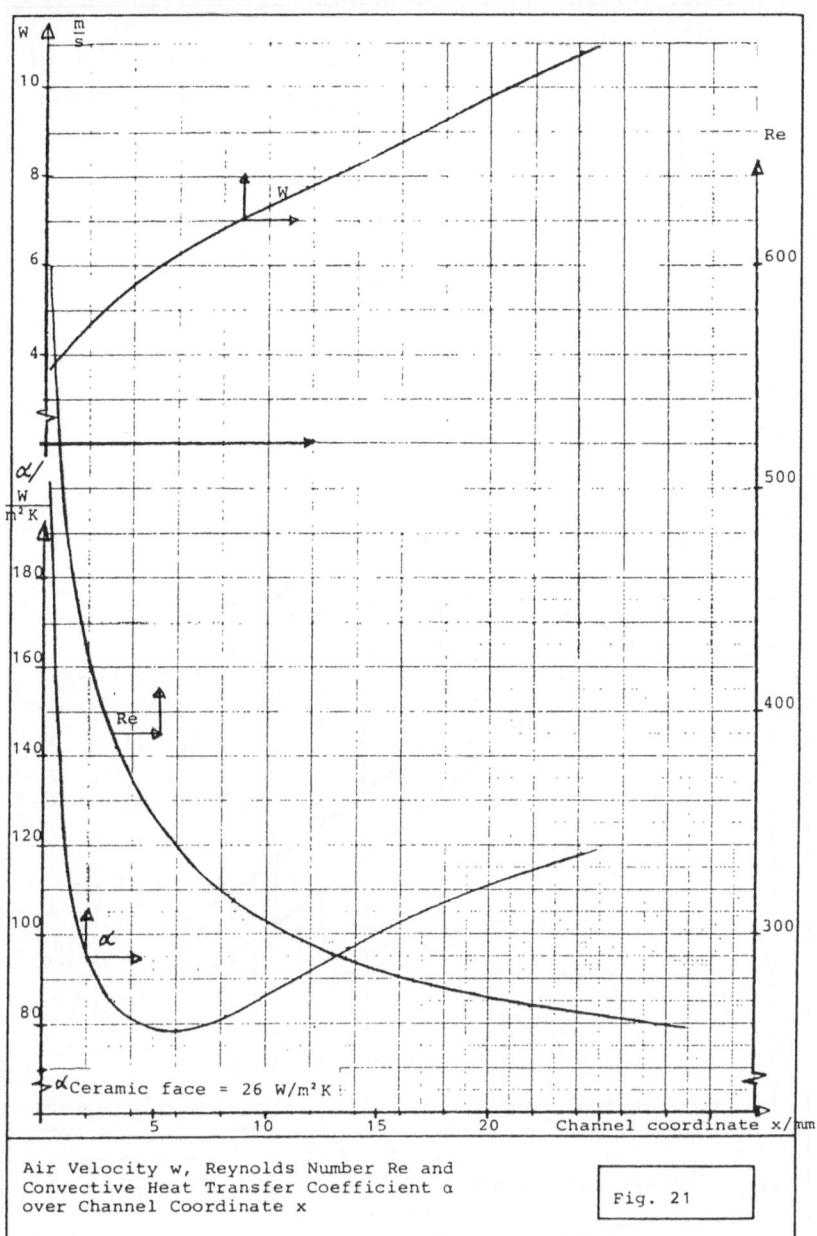

Air Velocity w, Reynolds Number Re and
Convective Heat Transfer Coefficient α
over Channel Coordinate x

Fig. 21

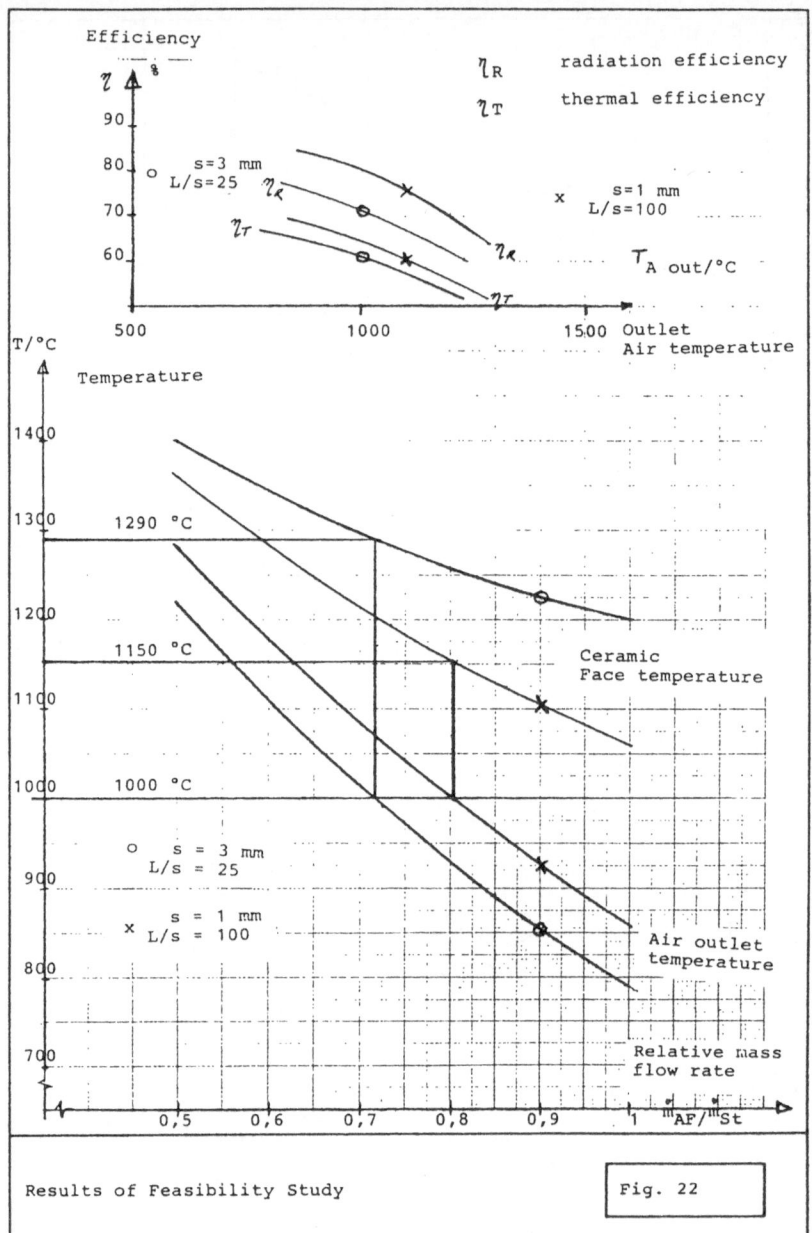

Results of Feasibility Study   Fig. 22

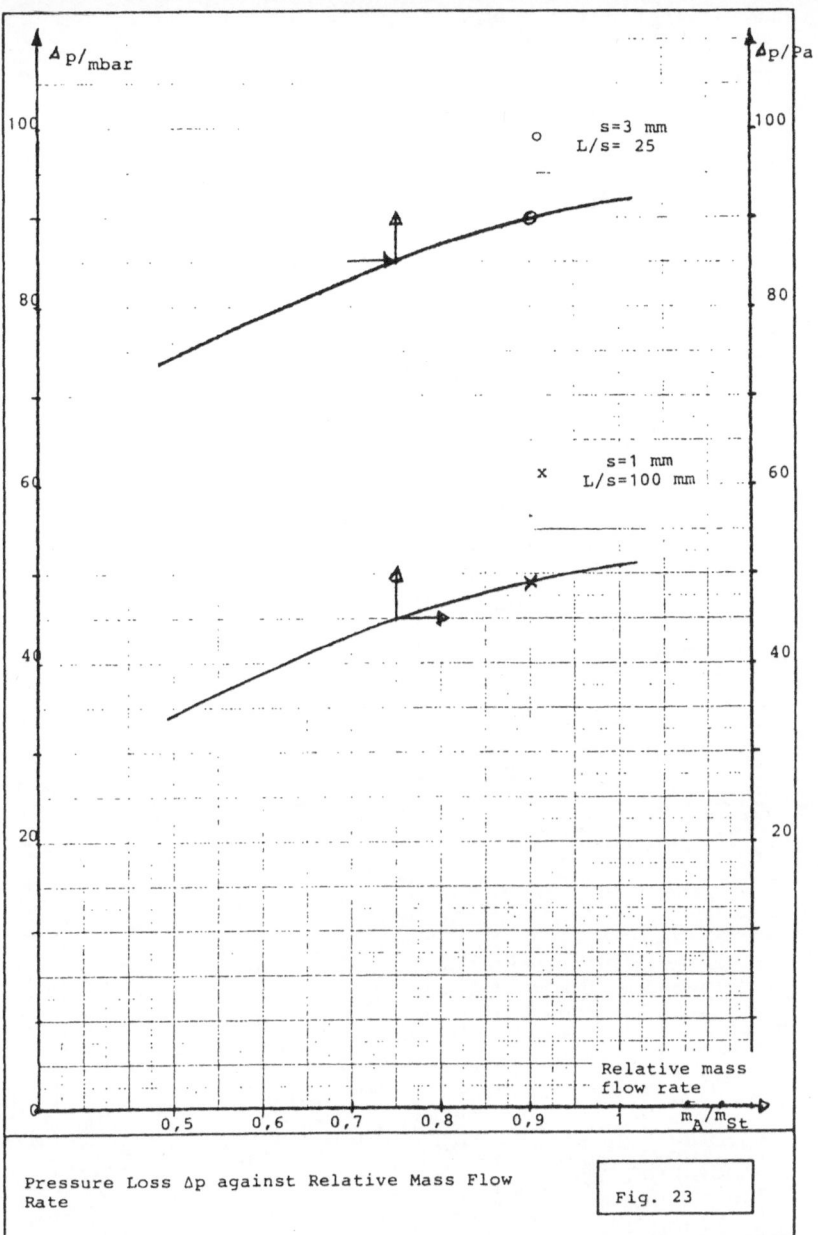

Pressure Loss Δp against Relative Mass Flow Rate

Fig. 23

INTERATOM

A MULTISTAGE STEAM REFORMER
UTILIZING SOLAR HEAT

W. JÄGER
U. LEUCHS
W. SIEBERT

INTERATOM, BERGISCH-GLADBACH

Contents

1   Introduction   61

2   Steam Reforming of Methane   65

3   Steam Reforming Plants   67

3.1 Conventional Plants   67
3.2 Convectionally Heated Steam Reformers   70
3.3 Multi-Stage Steam Reformer with Convectional Heating   73

4   Steam Reforming Plants with Solar Heating   78

4.1 Determination of the Design Data   78
4.2 Description of the Plant   81
4.3 Design of the Steam Reformer   88
4.4 Necessary Development Work   98
4.5 Characteristics of Operation   101
4.6 Costs of Manufacturing the Reformer   103

5   Literature   104

6   Figures   105

1. Introduction

Today a large amount of the required hydrogen or
synthesis gas (mixture of hydrogen and carbonmonoxide) is won by steam reforming of low hydrocarbons, especially methane. Hereby the mixture
of hydrocarbons and steam reacts under indirect
heating and by ways of a catalyst containing
nickel to hydrogen and carbonmonoxide as well
as carbondioxide. Hereat natural gas consisting
mainly of methane serves as feed gas. In order
to supply heat part of the feed gas is burned in
separate burners.
Depending on the feedback of the purge gas and the
heat recovery an amount of 25 to 30 % of the total
methane feed is necessary.

However, over the years the reserves of natural
gas will become scarcer. Therefore efforts are
being made to use methane only as raw material
instead of as fuel and to use other types of
energy to provide for the reaction heat and the
required heat for preheating.

For this, nuclear and solar energy can be used as
alternative energy sources. The study at hand
examines the use of solar energy. Already various
projects for steam reforming with solar energy
have been presented /1/, /2/. Also, power plants
(e. g. GAST) which generate electricity with solar
energy are beeing developed. Hereby temperatures of
800 ^{o}C by means of receivers made of metallic materials are reached. For higher temperatures, e. g.
up to 1000 ^{o}C, it is necessary to develop receivers
on a ceramic basis.

So far studies always dealt with conventional steam
reformers with reforming tubes of an outer diameter
ranging from 100 to 130 mm which are heated from the
outside and filled with catalyst. This reformer type
has proved to be successful in many plants with conventional radiation heated reformers and has also
been tested in pilot plants with convectional heating
by a helium flow. Hereby the helium had a temperature
of 950 ^{o}C.

In a concept for a reformer with solar energy using
modern receivers, which today can be laid out for
an output temperature of 800 ^{o}C, problems with conventional reformer tubes have arisen. If the required

reforming temperature for sufficient rates of
methane reforming is taken as a basis, then a
too small temperature difference will lead to
a too large tube surface. The then necessary
length of the reformer tubes would reach sizes
which would hardly permit building the tubes
and likewise, more problems would arise should
the number of parallel tubes be increased as
the velocity of the process gas would be too
slow. Both possibilities would require a much
too large and thus uneconomic catalyst volume
as a result of the large tube surface.

Therefore a new type of reactor, different to
the conventional reformers, is to be presented
and studied. For this type, a so-called multi-
stage reformer, the heat supply and catalyst
chamber are strictly seperated. Thereby the heat
can be supplied using thin-walled tubes with relatively small diameters, which have high heat
transfer coefficients as well as large heating
surfaces in relation to the volume of the heat
exchangers. Furthermore, the use of a hull-type
catalyst with narrow gas channels reduces the
required catalyst volume immensely and thus

offers an arrangement which allows a quick and easy change of the catalyst.

During the process in such a reformer the gas will be reformed adiabaticly in each stage of a multitude of catalyst stages and inbetween two stages then be heated to slightly higher temperatures than at the entrance of the previous stage. This results in a sawtooth formed curve of the reforming rate which accordingly runs along the curve of equilibrium.

The study at hand now deals with such a multi-stage reformer heated with 800 ^{o}C hot air.

2. Steam Reforming of Methane

The following equations describe the reforming of methane with steam :

$$CH_4 + H_2O = CO + 3H_2 - 206 \text{ KJ} \qquad (1)$$

$$CO + H_2O = CO_2 + H_2 + 41 \text{ KJ} \qquad (2)$$

The temperature dependence of both reactions is given as follows

$$K_{P1} = \frac{CO \cdot H_2^3 \cdot P^2}{CH_4 \cdot H_2O}$$

$$\ln K_{P1} = -11{,}692 - 0{,}1235 \cdot 10^{-2} \cdot T - 0{,}28578 \cdot 10^{-6} \cdot T^2 - 23270 \cdot \frac{1}{T} + 5{,}755 \cdot \ln T$$

$$K_{P2} = \frac{CO_2 \cdot H_2}{CO \cdot H_2O}$$

$$\ln K_{P2} = -13{,}268 - 0{,}31647 \cdot 10^{-3} \cdot T + 0{,}9296 \cdot 10^{-7} \cdot T^2 + 5320 \cdot \frac{1}{T} + 1{,}2293 \cdot \ln T$$

This shows that the very endothermic reaction 1 is superposed only by the weaker exothermic partial reaction 2.

Therefore the entire process is endothermic and thus
requires a considerable supply of heat.

Furthermore it can be concluded that high temperatures
at the end of the reaction and low process gas pressures
favor higher methane conversion. A higher ratio of steam
to methane promotes the reaction of carbonmonoxide and
water in the process gas to hydrogen. An insufficient
ratio of steam to methane will lead to a formation of
carbon according to the Boudouard reaction $2CO = CO_2 + C$

Molar ratios of steam to methane in the order of 2.5
to 5 are common.

3. Steam Reforming Plants

3.1 Conventional Plants

Today a great amount of the required hydrogen for
hydration processes or synthesis of ammonia and
synthesis gas for the synthesis of methanol or
the production of alcohols and aldehydes according
to the oxo-process and similar uses is produced
according to the steam reforming process described
in chapter 2. The process takes place in catalyst
filled tubes. During the process a heat supply is
necessary as the process is endothermic. The heat
is supplied by gasburners outside of the tubes.

Such a steam reformer consists of a fire chamber
which is a steel shell clad with several layers of
fire-bricks. Inside, the reformer tubes are spaced
in a wide pattern. The burners are situated inbetween the tubes on the top, bottom or sides. In
a ny case the heat is transferred solely by radiation.

The inner diameter of the reformer tubes ranges from
75 to 125 mm and the wall thickness from 15 to 20 mm.
The reformer tubes are made of heat resistant Cr-Ni-
cast steel.

Nickel coated aluminumoxide is used as catalyst.
The catalysts are commonly formed as Raschig rings
with a diameter of 10 to 16 mm which are packed
into the tubes irregularly. The lifetime of such
a catalyst is between 2 and 4 years.

Fig. 1 shows the scheme of a steam reformer with
burners situated at the top of the fire chamber /3/.
The feed gas mixture is preheated to 550 to 600 °C
and then distributed into the individual reformer
tubes. During the reforming the gas temperature increases continuously and reaches a temperature of about
800 °C at the end of the reaction. The process gas
is collected in a hot gas manifold and can then
serve as a supply for further uses.

In order to produce pure hydrogen the process gas
must be cooled down adequately so that the carbonmonoxide can be converted with steam into carbondioxide and hydrogen. During the synthesis of ammonia,
however, the gas flows from the primary steam reformer
into a secondary steam reformer in which the temperature rises to approximately 1000 °C on account of the
addition of air, which in turn results in an almost
complete reforming of the hydrocarbons. The gas then
also contains the required content of nitrogen.

This conversion is unnecessary for synthesis of
methanol or similar alcohols, since the carbonmonoxide in the gas is required for the synthesis.
Often the hydrogen content exceeds the carbonmonoxide content. In this case the hydrocarbons
are reformed with a mixture of carbondioxide and
steam instead of steam only. The process gas then
has a higher content of carbonmonoxide.

The steam reformer is heated by numerous burners
which are fired with methane (natural gas) lique-
fied gas or hydrocarbons up to benzene.

The waste heat of the process gas and flue gas
from all processes using a steam reformer is used
to preheat the feed gas and produce the required
process steam.

Fig. 1
conventional
steam reformer

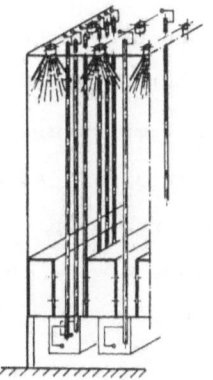

3.2 Convectionally Heated Steam Reformers

Among other energy sources, the PNP project
(Projekt Nukleare Prozeßwärme) studied steam
reforming with nuclear energy. Hereat the heat
is transferred from the reactor to the reformer
by a helium circuit. Contrary to conventional
plants the heat is transferred solely by convection instead of radiation. Since so far no
experience has been made with convectionally
heated reformer tubes a pilot plant called
EVA (Einzelspaltrohr-Versuchs-Anlage) was built.
The individual reformer tube had the same dimensions as the reformer tubes commonly used in conventional plants (an outer diameter of 130 mm, a
wall thickness of 20 mm and a heated length of
15 m) /4/. Therefore the results were also representative for the design of full-size plants.

Fig. 2 shows a flow sheet of the plant. Hereby
the helium is heated electrically instead of with
a nuclear reactor. A temperature diagram of the
test is shown in Fig. 3 /4/. As a result of the
extremely endothermic reaction, which starts very
quickly, the temperature of the process gas first
drops somewhat and then rises continuously.

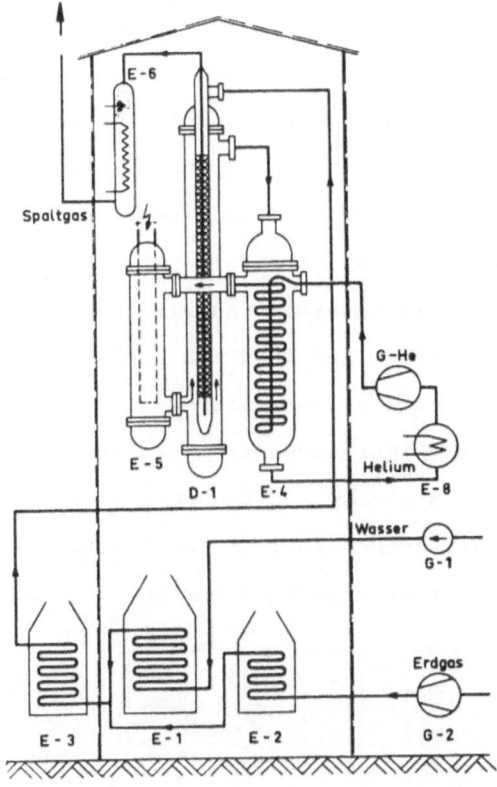

Spaltgas

E-6

G-He

Helium   E-8

Wasser   G-1

E-5   D-1   E-4

Erdgas   G-2

E-3   E-1   E-2

Tests have shown that the same surface power densities can be achieved with a mean temperature
difference of 150 °C as in conventional plants
with radiation heated reformers.

After several successful tests with the single reformer tube pilot plant a second larger pilot plant EVA 2 was built. The steam reformer had 30 reformer tubes and required 10 MW heat capacity. Two different bundles were tested, one tube bundle with baffle plates to produce a cross-counter-current-flow and one with a flow shield for each tube to produce a strict counter-current-flow. These tests likewise showed that entirely convectional heating of reformer tubes is possible and will have similar power surface densities and thus tube lengths as conventional radiation heated reformers, if helium with its excellent thermodynamic characteristics is used as heat transferring medium and if a sufficient temperature difference (helium temperature of 950 °C, process gas temperature of 800 °C) is available.

temperature

Fig. 3

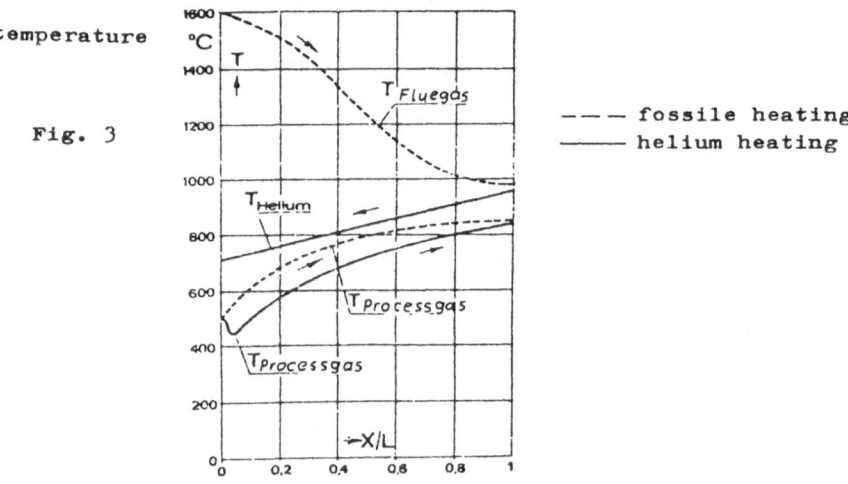

relative tube length

3.3 Multi-Stage Steam Reformer with Convectional Heating

In the previous chapter it has been proved possible
to heat steam reformers convectionally. In case of a
sufficient temperature difference between the heating
medium and the process gas the power surface density
is approximately the same as of radiation heated reformers. There is thus a certain relation between the
tube surface and the inner volume of the tube, respectively the catalyst volume.

If only a heating medium with lower temperatures
is available, the required tube surface will increase
immensely the lower the mean temperature difference
is. This results in very long reformer tubes or in
a larger number of parallel tubes if a certain length
limited by the possibility of construction has been
reached. With regard to the for the heat supply required tube surface the catalyst chamber will in each
case be much too large, thus also resulting in an un-
required amount of the expensive catalyst.

It would now be possible to lower the temperature of
the process gas in order to achieve a sufficient and
adequate mean temperature difference again. However,

this would have the disadvantage that the equilibrium would be further away from the hydrogen side so that less methane would be reformed. The process gas would then contain accordingly more and more unreformed methane, which would lead to difficulties in the following processes or to a large amount of circulating gas for all circuit processes.

It is thus the advantage of the multi-stage steam reformer developed by Interatom that the heat transfer surface and the catalyst chamber are separate and independent of one another. Therefore low heating media temperatures still allow process gas temperatures as high as possible, thus accepting a low mean temperature difference. Furthermore thin-walled tubes with small tube diameters can be used to transfer the heat. The overall volume and weight is small despite the large heat exchange surface.

Fig. 4 shows an outline of such a reformer. Inside a shell the catalyst is arranged in several layers. The heating tubes are grouped concentricly in an outer ring chamber. The feed gas flows through the first catalyst layer whereby part of the methane is converted with steam to hydrogen and carbonmonoxide and -dioxide.

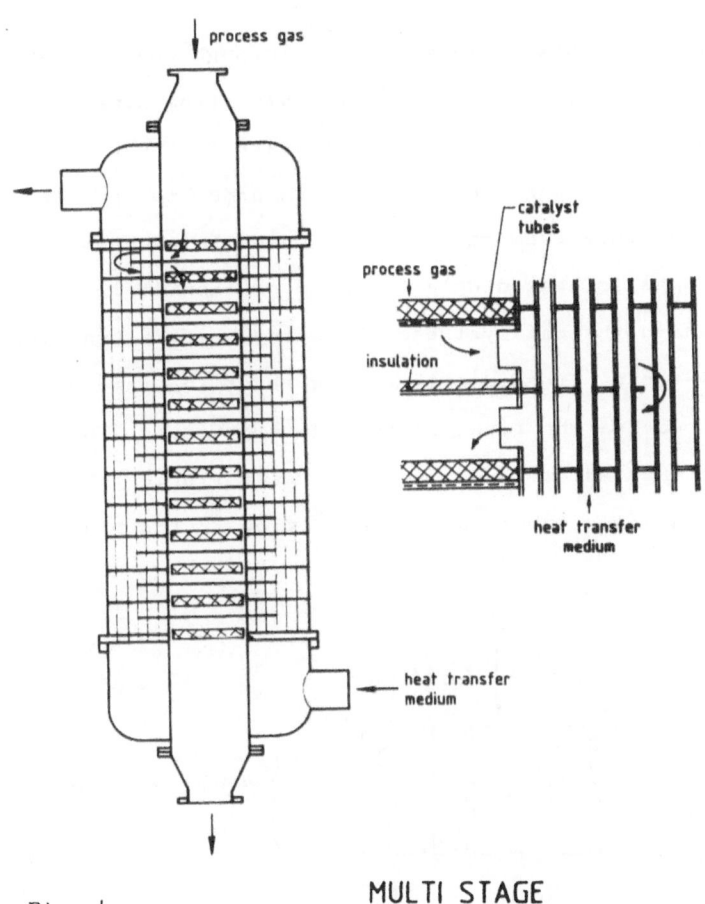

Fig. 4

MULTI STAGE
STEAM REFORMER

On account of the endothermic reaction the temperature
sinks somewhat. The gas then enters the outer chamber
where it is heated up by flowing round the heating tubes.
After reversing the gas flow in the tube chamber the gas

enters the next catalyst layer through openings in
the shell. This process is repeated from stage to
stage until the desired final state is reached.
20 to 25 stages are necessary in order to keep the
temperature from changing too much. In each stage
the gas is heated up a little more than it is cooled
down in the catalyst bed. Thus the temperature rises
saw-tooth   formed with rising reforming rate until
it reaches the final reforming temperature. The
heating medium is cooled down while flowing through
the heating tubes in a counter flow to the process
gas.

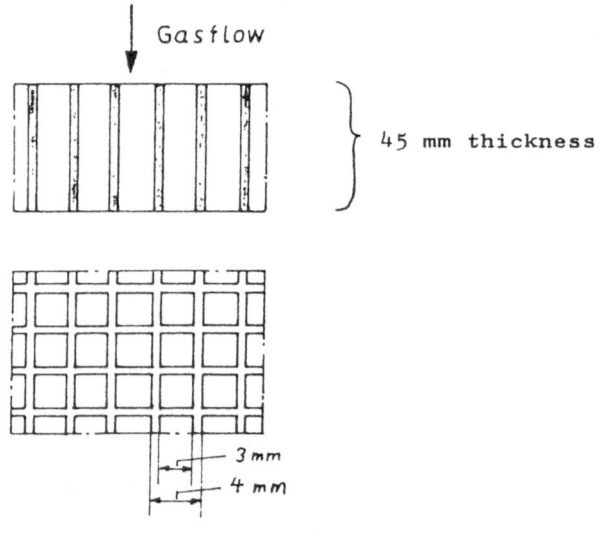

Fig. 5   catalyst

It is considered to use as a catalyst layer a very
finely meshed hull-type catalyst. The dimensions
of the catalyst are shown in Fig. 5. On account of
the large active surface the catalyst volume is by
far smaller than that of conventional dumped catalysts.

Furthermore it is comparatively very easy to change
the catalyst. The catalyst layer and the intermediate
plates can be mounted on rods and thus installed and
removed as a unit. The time to change the catalyst
lies in the range of only several hours compared to
several days or up to three weeks in conventional
plants.

The catalyst considered here has not to bear that
strong mechanical stresses during temperature changes
which occur in conventional catalyst filled tubes by
different thermal expansion of the tube- and catalyst
material. This effect may be a considerable additional
advantage of the hull-type catalyst in case of solar
application.

4. Steam Reforming Plants with Solar Heating

4.1 Determination of the Design Data

The receiver determines an air temperature of 800 °C. With regard to a sufficient temperature difference, the final reforming temperature would be 750 °C. The pressure should now be determined so as to achieve an adequate reforming rate of methane. Tab. 1 and Fig. 5 demonstrate the influence of the pressure on the composition of the process gas. Hereat following parameters were kept constant :

final reforming temperature	750 °C
equilibrium of methane reforming	740 °C
equilibrium of shift reaction	750 °C
ratio steam to methane	3

Table 1

P =		5	10	15	20
V =		5,7160	5,4484	5,2767	5,1578
C =	CH_4	2,484	5,062	6,854	8,164
	H_2	51,487	46,362	42,697	39,959
	CO	8,556	6,806	5,693	4,936
	CO_2	6,455	6,486	6,404	6,288
	H_2O	31,018	35,284	38,352	40,653
Q =		7957	6646	5814	5242

P = Final reforming pressure (bar)

V = Process gas volume per m^3 CH_4 feed (m^3/m^3)

C = Gascomposition (mol. %)

Q = Reaction heat (kJ/m_N3)

A final reforming pressure of 6 bar was chosen in order to obtain a sufficient reforming rate of methane and to compensate the lowered final reforming temperature.

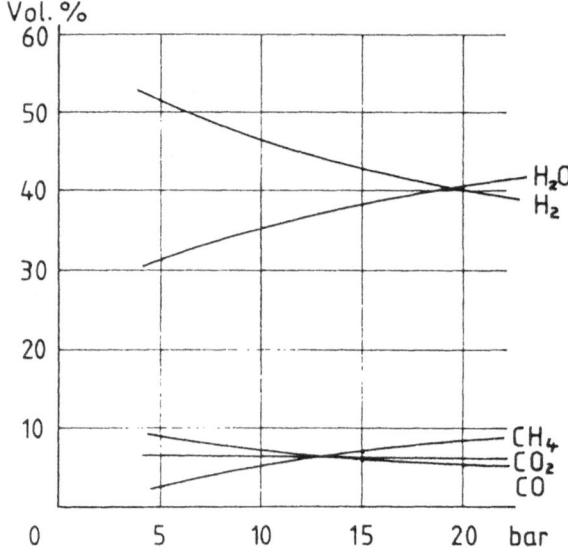

Gascomposition in relation to the pressure

Fig. 6

The plant is based on a capacity of 40 MW for the receiver.

4.2   Description of the Plant

In the previous chapter the receiver was said to
have a capacity of 40 MW. The plant was now designed so that the mentioned capacity of the receiver meets the heat requirements of the plant,
which are the requirements of the reformer and
the steam generator. Thus the reformer demands a
thermic capacity of 27,16 MW.

The process flow sheet of the designed plant is
shown in Fig. 7. The methane (or rather the natural gas) and the steam are mixed, in this case
in a ratio of 1:3 , and then heated to 600 $^{\circ}$C in
a heat exchanger while flowing counter-current to
the 750 $^{\circ}$C hot process gas coming from the reformer.
With this temperature the gas now enters the multi-
stage reformer which in the present case has 22 catalyst stages. In each stage the gas is converted
adiabaticly, which means that the temperature sinks
as a result of the endothermic reaction. Before
entering the next catalyst stage the gas passes
through the ring chamber where it is heated to a
higher temperature by the hot air flowing through
the heating tubes. Inbetween the catalyst stages

the gas is heated up more than it is cooled down
during the conversion, so that the temperature continuously rises saw-tooth formed until it reaches
the final reforming temperature. Fig. 8 shows the
temperature curve and Tab. 2 summarizes the main
process data of the reformer. The process gas exits
the reformer at 750 ^{o}C and is cooled down to 388 ^{o}C
in a heat exchanger used for preheating the feed gas.
In the following feed water preheater 2 the process
gas is cooled down further to 221 ^{o}C and in the feed-
water preheater 1 to 119 ^{o}C.

The 800 ^{o}C hot heating air coming from the receiver
enters the reformer where it supplies the heat for
the reforming process and thus cools down to 625 ^{o}C.
The air is then cooled down further to 335 ^{o}C in a
heat exchanger serving as air preheater and with this
temperature now enters the steam generator where it
is cooled down to 216 ^{o}C. In the following circulator
the air is first heated up to 250 ^{o}C and then heated
up further to 540 ^{o}C in the then following air preheater to thereafter enter the receiver at this
temperature. A fossile fired air preheater is installed parallel to the receiver which preheats
the air until the receiver has reached its full
capacity.

Table 2

Process Data of the Steamreformer

Feed gas:	Gasflow		12,093 m^3/s (i. N.)
	Gascomposition	CH_4	24,97 Vol. %
		N_2	0,11 Vol. %
		H_2O	74,92 Vol. %
	Pressure, Inlet-Reformer		6,8 bar
	Temperature, Inlet-Reformer		600 $^\circ$C
Process gas:	Gasflow		17,077 m^3/s (i. N.)
	Gascomposition	CH_4	3,09 Vol. %
		H_2	49,96 Vol. %
		CO	8,41 Vol. %
		CO_2	6,18 Vol. %
		H_2O	32,28 Vol. %
		N_2	0,08 Vol. %
	Pressure, Outlet-Reformer		6 bar
	Temperature, Outlet-Reformer		750 $^\circ$C

Thermal Power of the Reformer   27,156 MW

Heat transfer medium:

	Airflow	136 kg/s
	Inlet Temp. Ref.	800 $^\circ$C
	Outlet Temp. Ref.	625 $^\circ$C

The feed water is heated up to 90 $^{\circ}$C in the feed water
preheater 1 and then enters the deaerator where the
injection of steam leads to the deaeration and to a
rise of the feed water temperature up to 100 $^{\circ}$C. The
main feed water pump now pumps the feed water into
the feed water preheater 2 where the water is heated
to boiling temperature and 4,2 % of the water pre-
evaporized. Then the feed water is evaporized and
superheated to 220 $^{\circ}$C in the steam generator.

The process data of the steam reformer have been
summarized in Tab. 2. The other process data are
as follows:

Gas Preheater

feed gas: amount and composition acc. to Tab. 2

               inlet temperature   169 $^{\circ}$C

               inlet pressure   6,9 bar

               outlet temperature   600 $^{\circ}$C

               outlet pressure   6,8 bar

process gas: amount and composition acc. to Tab. 2

               inlet temperature   750 $^{\circ}$C

               inlet pressure   6 bar

               outlet temperature   380 $^{\circ}$C

               outlet pressure   5,9 bar

Feed Water Preheater 2

process gas: amount and composition acc. to Tab. 2

inlet temperature	388 °C
inlet pressure	5,9 bar
outlet temperature	221 °C
outlet pressure	5,8 bar

feed water: mass flow · · · · · · · · · · · · · 8,96 kg/s

inlet temperature	100 °C
inlet pressure	11 bar
outlet temperature (4,2 % pre-evaporized)	180 °C
outlet pressure	10 bar

Feed Water Preheater 1

process gas: amount and composition acc. to Tab. 2

inlet temperature	221 °C
inlet pressure	5,8 bar
outlet temperature	119 °C
outlet pressure	5,7 bar

feed water: mass flow · · · · · · · · · · · · · 8,8 kg/s

inlet temperature	20 °C
inlet pressure	2,2 bar
outlet temperature	90 °C
outlet pressure	1,2 bar

Steam Generator

heating air:	mass flow	136 kg/s
	inlet temperature	335 °C
	inlet pressure	5,9 bar
	outlet temperature	216 °C
	outlet pressure	5,8 bar
water/steam:	mass flow	8,96 kg/s
	inlet temperature (4,2 % evaporized)	180 °C
	inlet pressure	10 bar
	outlet temperature	220 °C
	outlet pressure	8 bar

Deaerator

feed water:	inflowing mass flow	8,8 kg/s
	inlet temperature	90 °C
	outflowing mass flow	8,96 kg/s
	outlet temperature	100 °C
deaeration steam:	mass flow	0,16 kg/s

Air Preheater

air flow:	primary and secondary	136 kg/s
primary side:	inlet temperature	625 $^\circ$C
	inlet pressure	6 bar
	outlet temperature	335 $^\circ$C
	outlet pressure	5,9 bar
secondary side:	inlet temperature	250 $^\circ$C
	inlet pressure	7 bar
	outlet temperature	540 $^\circ$C
	outlet pressure	6,9 bar

Receiver

thermic capacity	40 MW
air flow	136 kg/s
inlet temperature	540 $^\circ$C
outlet temperature	800 $^\circ$C

Air Heater

same thermic capacity and air
flow as receiver
maximum fuel gas consumption 1,19 m_N3/s

Air Circulator

air flow	136 kg/s
inlet pressure	5,8 bar
inlet temperature	216 $^\circ$C
outlet pressure	7 bar
outlet temperature	250 $^\circ$C
power consumption	4,9 MW

4.3 Design of the Steam Reformer

4.3.1 Description of the Design

General Lay-out

The multi-stage reformer is a cylindrically formed
pressure vessel. The front sides are sealed with heads.
One dished head is flanged. The cylindrical parts increase in diameter towards the dished heads. In this
enlargened part of the shell the inlet and outlet
nozzles for the process gas are arranged. The angular
position of the nozzles can be chosen so as to meet
the requirements of the plant to a great extent. The
inlet and outlet nozzles for the heating air are
positioned in the center of the dished heads. The first
and the last catalyst stage are positioned in the inlet and outlet nozzle of the processgas because the
reforming process should be started and finished by
a catalytical reaction and not by a heating step.

This type of multi-stage reformer is preferably positioned upright. A support cylinder with a mounting flange
connected to the upper enlargement of the shell can be
used to attach the reformer to the support construction.

In the following chapters it will be presumed that the
multi-stage reformer is arranged in an upright position.
Fig. 9 shows the general lay-out of the reformer, Fig.
10 and 11 details.

Inlet and Outlet of the Process Gas

The inlet nozzle (1) of the process gas is situated
in the vicinity of the upper enlargement of the shell.
The diameter of the connection depends on the required
size for the flanged catalyst layer (1.1) . The catalyst layer is installed in a holding ring matching the
flange series. An appropriate reducer is necessary to
connect the inlet duct for process gas. The nozzle is
attached to the ring shaped process gas distributor (2)
This annular channel is formed by the enlargened outer
shell and a dividing side (2.1) which is equivalent
to the extended outer shell.

The outlet nozzle (5) of the process gas is arranged
in the vicinity of the lower enlargement of the shell.
It is connected directly to the process gas collecting
chamber (4) which is formed by the part of the annular
channel not covered by the inlet (6) of the heating air.
The enlargement of the shell results from the required
space between the inner side of the shell and the floating tube sheet (8.2) . A catalyst layer (5.1) is installed on the flange side of the outlet nozzle and
determines just like for the inlet nozzle the dimensions
of the nozzle. Here too, the outlet duct for process gas
must be connected with an appropriate reducer.

Reaction Chamber

The reaction chamber $\left(3\right)$ is in the center of the multi-stage reformer. Except for the above two mentioned, all catalyst layers are arranged in the reaction chamber. The reaction chamber is formed by a perforated hull $\left(3.1\right)$ in the reaction chamber, the lower tube sheet $\left(8.2\right)$ and a cover plate $\left(3.2\right)$ whi.n covers an access opening in the upper tube plate.

The hull $\left(3.1\right)$ in the reaction chamber is attached to the upper tube plate $\left(8.1\right)$, whereas it is only guided at the lower tube plate $\left(8.2\right)$. The cross sectionsof the opening in the hull have been chosen as large as possible.

The honeycomb shaped catalyst layers $\left(3.3\right)$ are stacked. They are attached to the hull $\left(3.1\right)$ in the reaction chamber with a support construction. Separation plates $\left(3.4\right)$ separate the stages of the reformer.

The separation plates are also attached to the inner side of the hull in the reaction chamber. The junction is manufactured sufficiently tight. The dimensions of the heat exchanger segments determine the spacing of the separation plates.

Inlet and Outlet of the Heating Air

The inlet connection ⑥ for the heating air is at the lower dished head. The expansion joint ⑥.₁ is attached to the connection and the heating air distributor ⑦ which is welded to the lower floating tube sheet ⑧.₂ of the heat exchanger. The expansion joint is flanged to the inlet connection, the flange connection is accessible from the gas collecting chamber ④. The heating air distributor is formed as a dished head ⑦.₁ , the top end is closed by the floating tube sheet of the heat exchanger.

The heating air collecting chamber ⑨ is above the heat exchanger. It is formed by the upper clamped tube sheet ⑧.₁ and the upper dished head. The outlet nozzle ⑩ for the heating air is at the upper dished head and thus directly connected to the collecting chamber.

Heat Exchanger

The heat exchanger consists of the vertically arranged tubes, the two tube sheets and the baffle and separation plates. The outer surface of the heat exchanger is formed by the case, the inner surface by the hull in the reaction chamber.

The upper tube sheet (8.1) is flanged. It separates
in the upper zone of the reformer the process gas
from the heating air and is thus equipped with
appropriate seals. Furthermore this is the fixed
point of the whole heat exchanger including the
integrated reaction chamber. The force is introduced over the support cylinder to be carried by
the support construction.

The heat exchanger tubes (8.3) are welded to the
upper tube sheet.

In the center of the upper tube sheet is a circular
opening with a slightly larger diameter than the
catalyst layers. The catalyst layers and separation
plates can be removed through this opening to change
the catalyst. The opening is closed with a cover
plate (3.2) which is flanged to the upper tube
sheet and sealed.

The lower tube sheet (8.2) floats. It is connected
with the upper tube sheet by the heat exchanger
tubes welded to the upper tube sheet. The lower
side of the lower tube sheet is welded to the
heating air distributor (7).

Inbetween the tube sheets separation plates (8.4) are arranged horizontally. These subdivide the heat exchanger into individual stages. The height of the stages and thus the spacing of the separation plates depends on the required heat for each stage. Between every two separation plates of a stage is a baffle plate (8.5) which is necessary to re-direct the flow of the process gas.

4.3.2 Manner of Function

The process gas flows from the inlet duct through the inlet nozzle (1) into the multi-stage reformer, thereby passing through the catalyst (1.1) (first stage). The gas distributor (2) then leads the gas into the first stage of the heat exchanger (8) where it is heated up. Then the gas enters the reaction chamber (3) through the openings of the hull in the reaction chamber, flows through the catalyst (3.3) (second stage) and then, re-directed by the separation plates (3.4) , the gas flows back through the openings of the hull into the heat exchanger where again it is heated up. This process is repeated in the following stages in-cluding the twenty first stage. Then the process gas flows through the last stage of the heat exchanger into the gas collecting chamber (4) at the lower tube sheet and then through the catalyst (5.1) and the outlet nozzle (5) (twenty second stage) into the outlet duct.

The heating air enters the multi-stage reformer at the heating air nozzle. From the reformer the heating air flows through the heating air distributor (7), the heat exchanger (8) and the heating air collecting chamber (9), leaving the reformer at the heating air outlet nozzle (10).

4.3.3 Change of the Catalyst

The catalyst layers must be changed when the catalyst has lost its activity.

Following operations are therefore required (after the multi-stage reformer has been shut down and has cooled down) :

1) change of the catalyst layers in the reaction chamber

 - the outlet heating air ducts (an elbow type would be preferable) must be disconnected and removed from the heating air nozzle (10)

- the insulation ⑩.1 around the nozzles must be removed

- the cover plate ③.2 must be disconnected and removed

- the catalyst layers ③.3 and appropriate separation plates ③.4 must be taken apart stage after stage and removed

The reaction chamber can then be reloaded. In opposite succession, the catalyst layers are then installed and the openings closed. The tightness of the flange connections must be checked.

2) change of the catalyst layers ①.1 and ⑤.1 in the nozzles

- the bolts of the flanges must be released and the flanges jacked with jack screws

- the catalyst layer and the holding ring must be removed

- the new catalyst layer must be installed

- the flange connection must be torqued, bolted and sealed

4.3.4   Technical Data of a Multi-Stage Reformer

thermal power		27,2 MW
total length		15000 mm
length of the heat exchanger between the tube sheets		12000 mm
diameter at the gas distributor		2900 mm
diameter at the gas collecting chamber		2800 mm
feed-gas inlet :	dimension	DN 700
	temperature	600 °C
	pressure	6,8 bar
reaction chamber:	diameter (inside)	1200 mm
	length	12000 mm
	number of catalyst layers	20 + 2
process gas outlet :	dimension	DN 700
	temperature	750 °C
	pressure	6,0 bar
heating air inlet :	dimension	DN 1200
	temperature	800 °C
	pressure	6,8 bar
heat exchanger :	active length	12000 mm
	number of tubes	1300
	spacing	55 mm
	tube dimensions	42,4 x 2,6 mm

```
heating
air outlet :        dimension              DN 1200

                    temperature            625 °C

                    pressure               6,0 bar


weight              about                  130000 kg
```

4.4   Necessary Development Work

The multi-stage steam reformer is a new concept for
a reformer. Therefore certain work is still necessary
before a first full-scale plant can be built. This
development should be conducted in 2 stages.

Stage 1:   Development of the Catalyst
The considered honeycomb shaped catalyst for the reformer has not yet been installed in a steam reforming
plant. This type of catalyst has already been used for
other catalytic processes and proved to be very advantageous on account of its large surface. For instance,
in several cases it was possible to reduce the required
catalyst volume to 20 % of the required amount for
dumped catalysts. Already the carrier on the basis
of aluminumoxide is produced in large amounts and can
be procured. The Südchemie Company in Munich has offered
to produce and test this catalyst. First the carrier
must be soaked in a solution containing nickel to a
content of $\sim$ 25 % NiO. Short time testing must prove
following qualities of this catalyst:

- high activity
- easy to reduce
- high mechanical durability

The Südchemie Comp. has estimated for the development
of the impregnated catalyst and the then necessary
tests to prove the above mentioned qualities:

- duration :   one month
- costs :   about 20.000,-- DM (for three testruns)

Stage 2: Construction and Operation of a Pilot Plant
The construction of a pilot plant should now demonstrate
the following:

- possibility to construct a multi-stage reformer
- operation of the reformer with adiabatic conversion in each stage
- heat supply
- long term   characteristics of the catalyst
- characteristics of operation for varying capacity
- required minimum power input to keep up the
  reforming process
- maximum span of time in which the reformer can
  stay shut down and directly resume operation
  without going through the complete process of
  shutting down and starting up
- starting up and shutting down behaviour

A reformer capacity of 1 MW would be representative. This reformer would have a flow of 400 m_N^3/h methane and convert 330 m_N^3/h methane. The inner chamber for the catalyst would have a diameter of 225 mm.

This pilot plant could be erected on the Almeria site where the methane supply for the long term tests would be problematic, or here in Germany where solar conditions would have to be simulated.

The costs of such a pilot test here in Germany are roughly estimated to

investment costs:	4,5 Mio DM
yearly operation costs:	7,0 Mio DM

Should the financial means at the beginning of stage 2 be limited it would also be possible to test a smaller reformer e. g. with a heat capacity of 0,2 MW, in a fossile fired plant and then test the 1 MW pilot plant later directly in a solar plant.

4.5 Characteristics of Operation

The reformer is started up and shut down just like
for conventional catalysts by reducing the oxidized
catalyst. As for all catalysts on a ceramic basis
the rate of heating up is not allowed to exceed
1 K/min. Considering the relatively long duration
for starting up and shutting down the plant should
thus not be shut down overnight but pursue operating
with fossile firing.

Here are exist the following two possibilities:

- the plant continuously operates with a full
  capacity, the heat not supplied by solar
  energy being supplied by fossile fuels
  (see Fig. 12)
- the capacity of the plant is reduced to $\sim$ 30 %
  according to the heat supplied by solar energy
  and operation then continued with fossile firing
  until the heat supplied by solar energy increases,
  in which case the capacity is then accordingly increased to full load (see Fig. 13)

The second possibility requires less fossile fuels, the plant is however operated at full capacity only for a short time per day. Whichever possibility is chosen depends e. g. on the costs of fuels or whether or not larger amounts of waste gas are to be fired at night. Basicly the previously described plant can be operated employing either possibility or any state inbetween.

4.6 Costs of Manufacturing the Reformer $(27,2 \text{ MW}_{th})$

In a first rough estimation the costs of constructing
the multi-stage reformer amount to:

Reformer   7 Mio DM

Catalyst   130 000 DM

Literature

/1/   Lurgi GmbH Frankfurt/Main
        Study on steamreforming of methane
        utilizing solar heat, March '84

/2/   Lurgi GmbH Frankfurt/Main,
        MAN Technologie GmbH München
        Study "Comparative Investigations and
        ratings of different solar systems using
        tubular steam reformers for grass root
        chemical complexes".

/3/   M.Apple, H. Gössling
        "Herstellung von Synthesegas nach dem
        Steamreforming Verfahren"
        Chemiker Zeitung 96 (1972) S. 135 - 153

/4/   R. Harth, B. Höhlein
        "EVA - Experimente zur Einkopplung von
        nuklearer Wärme in chemische Prozesse"
        Jahresbericht 1972 der Kernforschungs-
        anlage Jülich GmbH, S. 13 - 16

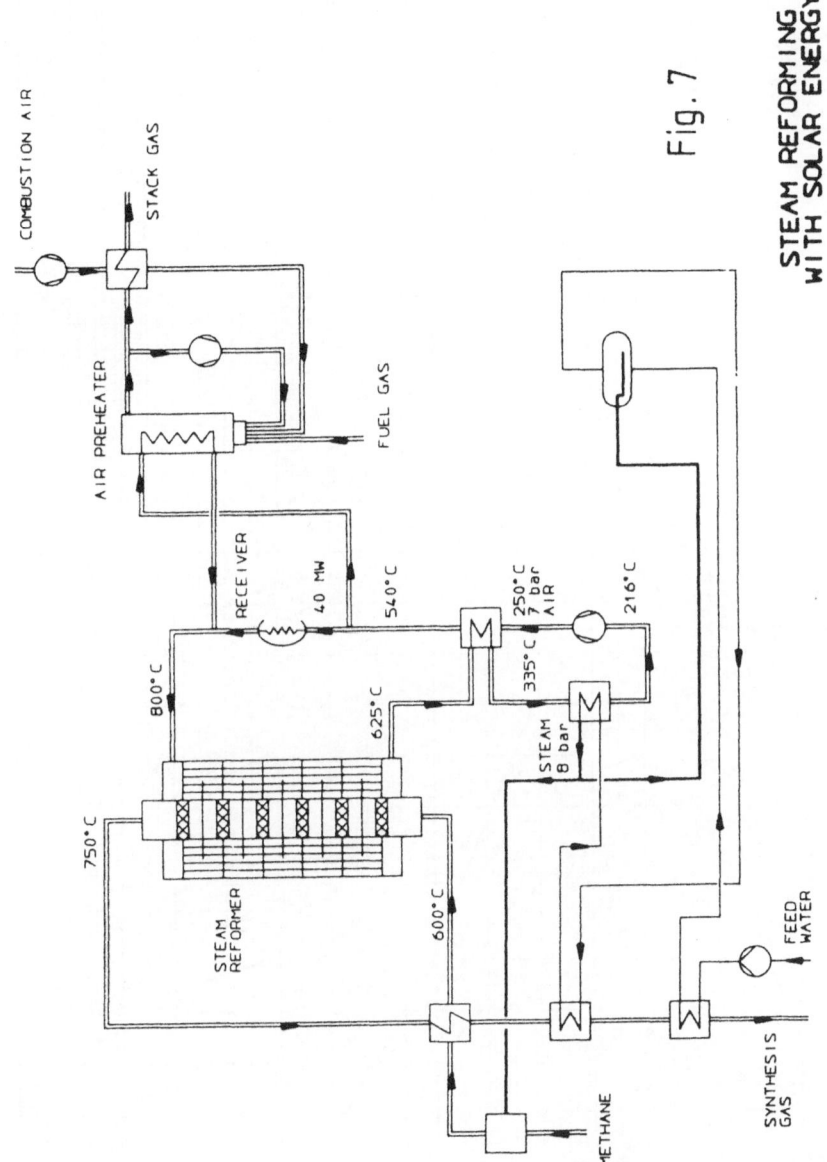

COMBUSTION AIR

STACK GAS

AIR PREHEATER

FUEL GAS

RECEIVER

40 MW

540° C

250° C
7 bar
AIR

216° C

800° C

625° C

335° C

750° C

600° C

STEAM
8 bar

STEAM
REFORMER

FEED
WATER

METHANE

SYNTHESIS
GAS

Fig. 7

STEAM REFORMING
WITH SOLAR ENERGY

- 105 -

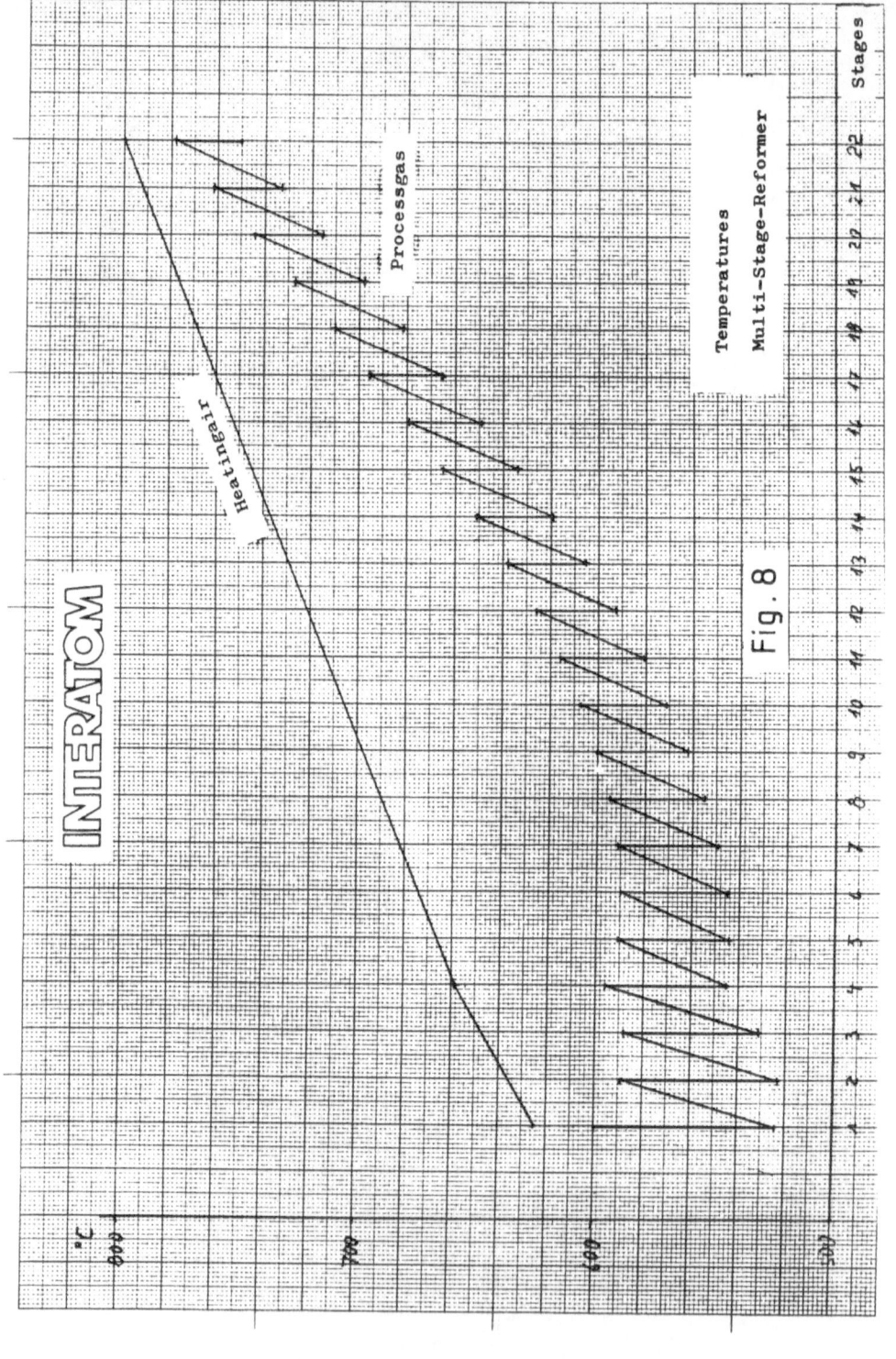

INTERATOM

Heatingair

Processgas

Temperatures
Multi-Stage-Reformer

Fig. 8

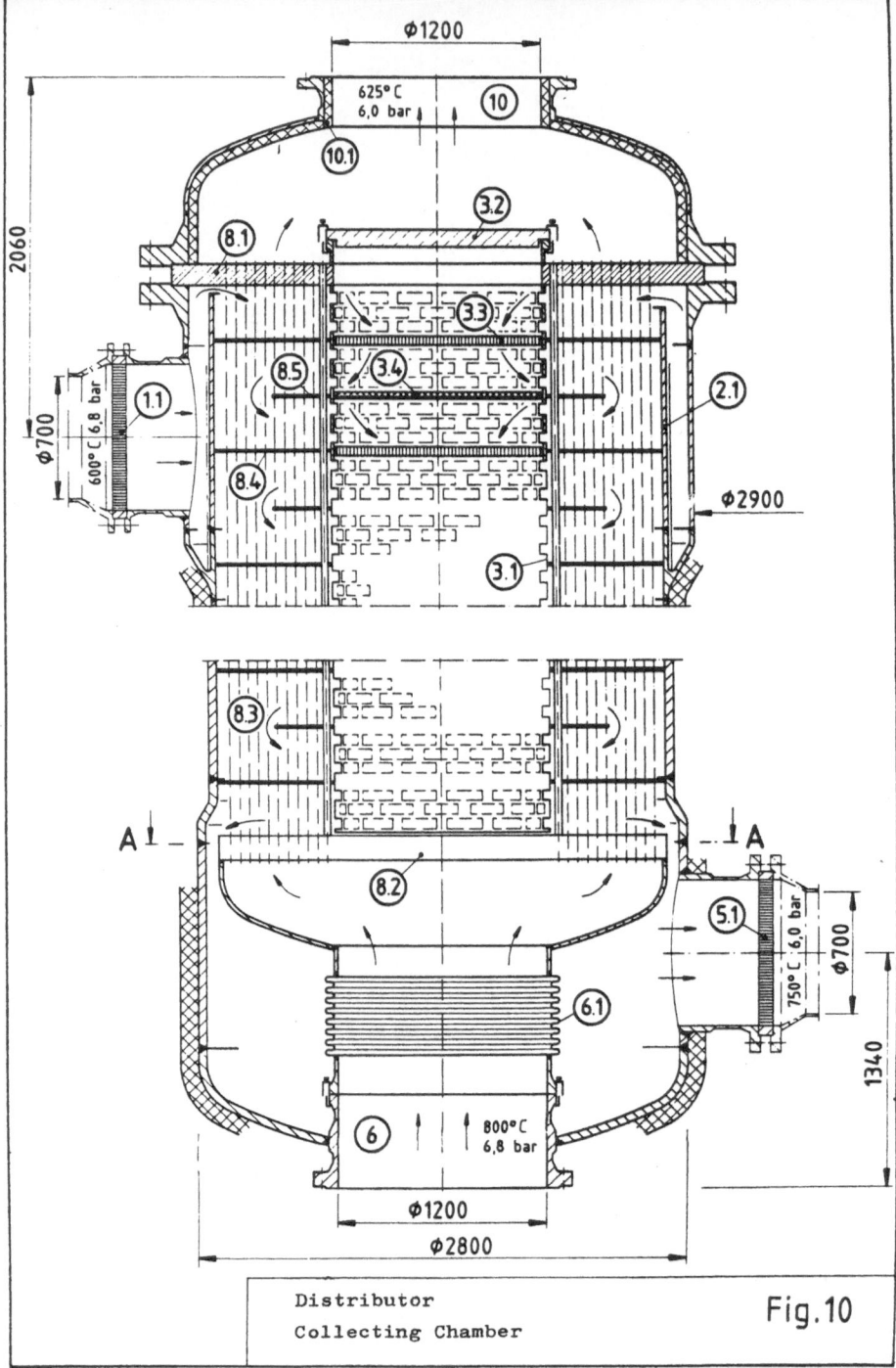

Distributor
Collecting Chamber

Fig.10

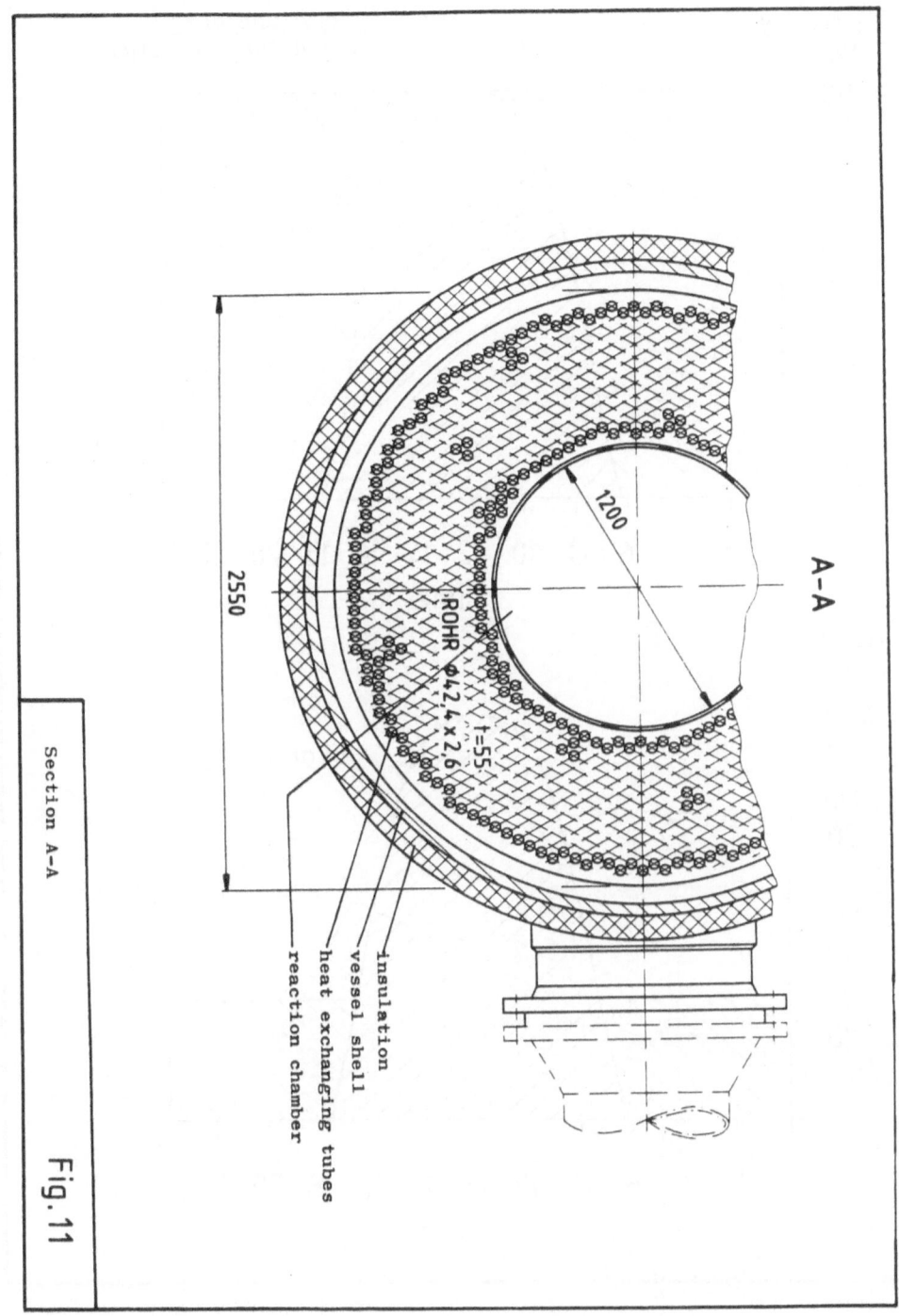

Section A-A

Fig. 11

A-A

2550

1200

t=55
ROHR Φ42,4 x 2,6

insulation
vessel shell
heat exchanging tubes
reaction chamber

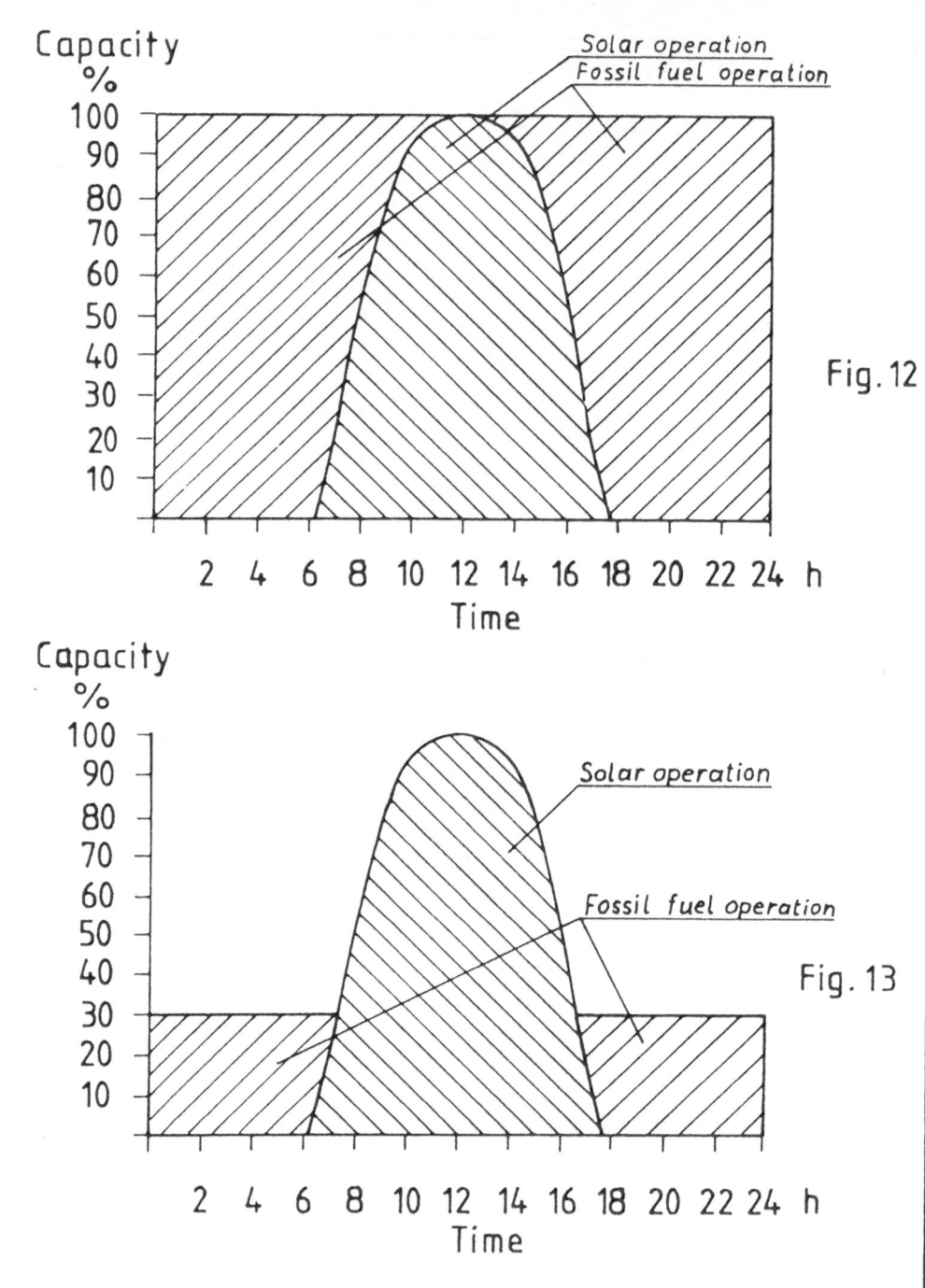

Capacity %

Solar operation
Fossil fuel operation

Fig. 12

Time

Capacity %

Solar operation

Fossil fuel operation

Fig. 13

Time

DIDIER-WERKE AG

DIDIER

LAYOUT OF HIGH TEMPERATURE
SOLID HEAT STORAGES

H. KALFA
C. STREUBER

DIDIER, WIESBADEN

Contents

1   Introduction   115

2   Tasks of the Study   115

3   General Informations Concerning Heat Storages   116

4   Test Storage in Germany (Stuttgart)   117

4.01   Test Storage Design   117
4.02   Components for the Test Storage   119
4.03   Estimated Costs for the Test Storage   122

5   Pilot Heat Storage in Spain (Almeria)   123

5.01   Conception of the Pilot Storage Plant   123
5.01.01 Number of Storages   123
5.01.02 Arrangement of the Storages   124

5.02   Loading and Unloading Cycle   125

5.03   Conditions for Design   126

5.04   Performance of Storages and Consumer   127

5.05   Design of the Storage and Lining   128

5.06	Connecting Mains	129
5.06.01	Receiver/Mushroom Valve Near Storage	129
5.06.02	Valve/Storage	130
5.06.03	Storage/Blower/Receiver	130
5.06.04	Storage/Stack	131
5.06.05	Cold Blast (Pressure) /Storage	131
5.06.06	Storage/Consumer	132
5.06.07	Cold Blast Mixing	133
5.07	Design of the Components	133
5.07.01	Pressure Units	133
5.07.02	Consumer	136
5.07.03	Blower (pressureless cycle)	137
5.07.04	Heating System	137
5.07.05	Stack	138
5.07.06	Valves	139
5.08	Energy Consumptions	140
5.08.01	Electrical Power Required	140
5.08.02	Water Consumption for Valves	141
5.08.03	Electrical Energy Consumption	141
5.08.04	Heat Losses	142
5.08.05	Temperatures in the Storage	142
5.08.06	Temperatures in a Connecting Main	143
5.08.07	Heating Up the Storages	143
5.09	Operation in Case of Disturbances	144
5.10	Estimated Costs (Pilot Storage Plant)	145
6	Commercial Heat Storage	146
6.01	Design of One Commercial Heat Storage	146
6.02	Design of a Commercial Heat Storage Plant	147
6.03	Operation of a Commercial Storage Plant	148
6.04	Temperatures in the Storage	148
6.05	Performance of the C S P	149
6.06	Estimated Costs of a "C S P" (3 Storages)	151
7	Figures	152
8	Drawings	206

1   INTRODUCTION

Blast furnace stoves are proven components for high temperature storage applications in steel plants. When storing heat energy, not only the energy itself but also the quality of the energy (high temperature) is important. The higher the temperature of the medium the higher the quality of the energy. As illustrated by the hot blast stove. it is indeed possible to store energy at a high temperature. It is therefore possible to operate a high temperature preocess.

The sun shines discontinuously. With the storage of heat it is possible to operate a continuous thermal process with the discontinuous energy by the sun shining on a receiver.

At the beginning of 1986 DIDIER carried out a preliminary study for a heat storage test plant. This present study was carried out after discussion of the results.

The study firstly generally discusses existing high temperature storages. This is followed by designs for temperature storages (storage plants).

2   TASKS OF THE STUDY

This study incorporates three main subjects:

- heat storage material test unit in Stuttgart,

- pilot heat storage plant in Almeria.

- commercial heat storage.

The study includes the design of the test unit and the pilot plant as well as the design of one storage component for the commercial storage. The operation of a commercial storage plant is also given. The performance of a 3-storage plant is given in relationship to the consumer power and the required efficiency of the heliostat field.

For the three subtasks an estimated price is given. The needed components are listed.

High temperature heat storages have an important role in production of both glass and iron. In blast furnace operation low cost fuels with low calorific values are burned in the hot blast stoves to preheat the combustion air for the blast furnace. High temperature and high pressure of the blast is possible. Hot blast stoves preheat up to 450,000 m3/h of air to temperatures up to 1400 deg C at pressures of up to 6 bar. The main elements are the checker work for the heat storage and the combustion chamber.

Such heat storages operate as a regenerator. A heat exchanger not yet is available in the given range of temperature and pressure.

With the regenerator it is possible to store heat (with high temperature and quality) without pressure. The unloading of the storage is possible with high pressure. Therefore a process (chemical or heat) can operate with high pressure and high temperature. Loading is normally effected by burning gas. In this study the loading is effected by the sun with a receiver.

Figure 1 shows a hot blast stove with internal combustion chamber. A storage for solar energy can be built to similar design but without a combustion chamber. The checker work inside the stove is the most important item in the heat storage. For high temperature the checker work is of refractory material of different qualities with different bulk densities. The checker work is built with special checker bricks. Figure 2 shows an example of a checker brick made by DIDIER for hot blast stoves. The air or the waste gas passes through the channels in the bricks.

A normal stove plant in the iron and steel industry has two, three or four stoves (storages). The minimum is two storages because at any given time one of the two stoves must be unloading. After changing over the unloaded stove again goes on load. The time for change over is dependent on the pressure of the blast and the internal volume of the storage. Figure 3 shows the change over from one stove to another as well as the instrumentation.

Figure 4 gives typical temperature curves in the checker work are shown. These temperature curves are applicable over the height of the checker work in every heat storage of a similar design.

A typical hot blast outlet of a hot blast stove is shown in figure 5. Figure 6 gives a detail of an overhanging dome on a special hot blast stove. These two details are shown because similar designs are applicable to the heat storages in this study.

4   TEST STORAGE IN GERMANY (Stuttgart)

4.01   TEST STORAGE DESIGN

The test storage must allow for variable modes of operation and peripheral conditions, more so than are required in the pilot or commercial storage plant. The following conditions must be possible:

- high pressure

- low pressure

- different temperatures

- high temperatures also at the bottom of the checker work (700 deg C)

- changing of the checker bricks must be very simple

- time for loading and unloading must be variable

- use of special checkers (hybrid material).

The following data are specified:

maximum temperature at the top   1300 deg C

maximum temperature at the bottom   700 deg C

maximum pressure when unloading   20 bar

pressureless loading with a burner not with the sun.

There are different ways of changing the checker work itself. The best way is to lift the checkers into the storage from a basement.

This enables the best possible closing of the storage in order to withstand the maximum pressure of 20 bar. When changing of the checkers will be done under the storage the height is limited.

The design of the checker is as follwos:

diameter of checker   500 mm

height of checker work   2000 mm

weight of checkers   700 kg

free cross-sectional area   40 %

The total height of the storage itself is 3,75 m, with a 3,5 m high space underneath it for changing the checkers. This space can be below ground level.

A drawing of the storage is given in figure 7.

The lining of the storage wall (insulation) is as follows:

refractory castables   160 mm

insulation castables   190 mm

steel   25 mm

The dome of the storage is insulated with refractory and insulation bricks.

The checker support must be a special alloy because the maximum temperature is 700 deg C.

With the given design the energy difference between load and unload is in the range of 20 kWh. The value is dependent on the loading / unloading time, the cycle, the temperatures (on top, at the bottom, when loading, when unloading etc.).

The other necessary components are shown in figure 8

COMPONENTS FOR THE TEST STORAGE

This chapter sets out a list of the components required. See chapter 4.01 for the storage itself. The medium is atmospheric air or waste gas from the burner for heating up.

burner and combustion chamber

performance	30 kW
maximum temperature	1300 deg C
minimum temperature	100 deg C (for drying)
pressure	1,01 bar

connecting main burner / storage

internal diameter	300 mm
external diameter	800 mm
insulation with castables	
design temperature	1300 deg C
design pressure	20 bar

connecting main storage / consumer

internal diameter	300 mm
external diameter	800 mm
insulation with castables	
design temperature	1300 deg C
design pressure	20 bar

connecting main consumer / storage

    internal diameter   100 mm

    external temperature   700 deg C

    design pressure   20 bar

other lower temperature connecting mains

    design temperature   700 deg C

    design pressure   20 bar

consumer

    heat exchanger

for high temperatures (1000 deg C) the hot air from the storage is firstly mixed to produce a lower temperature of 800 deg C and then cooled down to the inlet temperature at the lower end of the checker work (200 deg C).

booster

    design pressure   20 bar

    max. flow rate   200 Nm3/h

    pressure difference   10 mbar

compressor

for filling the storage to a max. 20 bar.

heat exchanger in the waste gas flow

This heat exchanger is only for cooling the waste gases from 700 deg C to about 200 deg C.

silencer

A silencer is required for changing from high pressure
unloading to loading. This restricts the noise level to
max. 55 dB(A) during the changeover period.

valves

2 cooled valves, 1300 deg C / pressure 20 bar

3 cooled valves, 700 deg C / pressure 20 bar

1 valve behind compressed air container

4.03   ESTIMATED COSTS FOR THE TEST STORAGE

The estimated costs for the test storage are total
2 950 000 DM.

The following items are not included in the price:
foundation
anchoring of the storage
heating media (oil or gas, electricity, etc.)

The costs are estimated costs of today.

5.01   CONCEPTION_OF_THE_PILOT_STORAGE_PLANT

5.01.01   NUMBER_OF_STORAGES

A plant was firstly designed with a big storage and 3
regenerators. Two alternatives of this design were discussed:

    - the large storage operating at 20 bar over the
      whole time,

        or

    - the large storage always operating at atmospheric
      pressure

The first alternative is shown in figure_9, the largest
storage is under pressure, the shell is extremly thick. The
heat losses are high because the heat is transported from the
receiver to the regenerator, thence to the storage and then to
the consumer. If the heat is only transported once the thermal
losses are then lower.

If the larger storage is operated without pressure (see
figure_10) the heat is also transported from one regenerator
to the other before it arrives at the consumer.

The best way to save energy and to have a normal thickness of
the shell is to have 3 regenerators of one size which are also
storages. A plant with only 2 regenerators isn't possible
because at any given time 1 storage/regenerator must be
loading and 1 unloading. The time for changeover from loading
to unloading is in the region of 20 to 30 minutes and
therefore 3 regenerators are needed. The time for changeover
is dependent on the pressure (20 bar) and the internal volume
of the storage.

The following advantages accrue for the design with 3
regenerators / storages instead of 1 storage and 3
regenerators:

- lower heat losses over the shell,

- shorter connecting mains,

- smaller volume,

- shorter connecting mains, smaller volumes and smaller surfaces also mean lower costs.

Figure 11 shows a more detailed sketch of the designed pilot plant. The components of the plant are given in the following chapters.

5.01.02   ARRANGEMENT OF THE STORAGES

For the purposes of the study the storages were arranged in line. A sketch of the 3 storages and the receiver tower is shown in figure 12. A detailed drawing of the storages and the components (see also figure 11) is given in drawing no. 86087 A. Figure 13 shows the situation in Almeria Plataforma Solar with the storages near the existing receiver tower. It seems to be possible that the three storages can be built up between the existing receiver tower and the heliostat assembly hall. Some components may stand in this hall.

Another arrangement is in form of a triangle. A sketch is given in drawing 86090. The receiver in this arrangement is on top of the middle of the storage triangle. This arrangement was not implemented in this study because:

- the connecting mains (hot) are longer than in the other arrangement,

- a reduction of heat losses isn't possible,

- a chamber must be built in the main from receiver,

- the connecting mains (cold) are longer,

- the compensation of the mains is a greater problem, because one of the storages on the triangle is on pressure, the other are not. In a row arrangement the compensation is easier.

5.02   <u>LOADING AND UNLOADING CYCLE</u>

This chapter shows the operation of the storage plant and especially the changeover from one to another storage.

The heating system provides for the initial heating of the storage plant, heating up after a break in the tests or for maintaining the storages and connecting mains at operating temperature. <u>Figure 14</u> shows this procedure.

The next 4 figures show one cycle.

<u>Figure 15</u> :

    - storage 1 and 2 loading,

    - storage 3 unloading,

    - the storages that are in the loading circuit are at atmospheric pressure,

    - the pressure in the unloading circuit is 20 bar.

A changeover for unloading from storage 3 to storage 2 will now be described.

<u>Figure 16</u> :

    - storage 1 still loading,

    - storage 2 cut off from loading and will be filled with air from the compressor up to 20 bar,

    - storage 3 still unloading.

When the pressure in storage 2 reaches 20 bar it is cut off from the compressor. The unloading circuit changes from storage 3 to storage 2. The next event is shown in:

<u>Figure 17</u> :

  - storage 1 still loading,

  - storage 2 unloading,

  - the pressure in storage 3 is initially 20 bar, the air expands through the silencer until it reaches ambient pressure.

The time to expand the air from 20 bar to ambient is shown in <u>figure 18</u>.

After storage 3 has reached ambient pressure it can then be loaded from the receiver.

<u>Figure 19</u> :

  - storage 1 and 3 loading,

  - storage 2 unloading,

  - the rest is similar to figure 15.

5.03   <u>CONDITIONS FOR DESIGN</u>

The circulating medium is atmospheric air.

Heat losses of the storages and connecting mains are related to:

        an ambient temperature of   20 deg C

        at atmospheric wind velocity of 4 m/s

The temperature at the inner surface of a lining is assumed as the operating temperature. When a correct heat transfer coefficient is calculated the heat losses are a somewhat lower, but the given value is the maximum value with the conditions given for the surrounding.

As discussed with the DFVLR the following design for the heat
storages and the consumer was chosen:

        consumer   300 kW thermal

        temperature hot   1000 deg C

        temperature cold   50 deg C

        volume air to consumer 800 Nm3/h

        pressure   20 bar absolute

Figure 20 shows the performance in relationship to the
temperature difference at the consumer.

The following performance of the receiver is assumed:
(figure 21)

        P(t) = 0,25 * PR + 0,75 * PR * f(t)

        f(t) = cos ((t/tD - 0,5) * PI)   0 < t < tD
        with
              tD = 8 h for the date 21.12.
              tD = 10 h for the date 21.03.
              tD = 12 h for the date 21.06.

PR is the maximum performance of the receiver (peak). The
temperature at the receiver outlet is a minimum of 1200 deg C.
The storages and the connecting mains are designed for a
maximum temperature of 1400 deg C.

The performance of the receiver itself can be calculated for
the dates (21.3./21.6./21.12.) when the heat losses of the
components are calculated.

The cycle for the storage operation was given in chapter 5.02.
Figure 22 shows how each storage / regenerator operates over
one day. The flow rates are calculated in accordance with the
times assumed in this figure. It can be seen that one
regenerator is always unloading. Each regenerator will be
loaded over the day (e.g. 21.06.) with the same amount of
energy. The peaks in the figure are those times when a
regenerator is changed from loading to unloading. If the
curves are summarized there is one constant line for unloading
and one curve similar to that in figure 21.

An alternative operation cycle is given in figure 23. The number of changovers is smaller over the day. Tests will prove which alternative has the better performance.

5.05   DESIGN OF THE STORAGE AND LINING

The checker work for the design case is calculated as follows (with a computer program from DIDIER) :

        weight of the checkers   approx. 50 t

        height of the checker work   9 m

        diameter of the checker work   2,25 m

        free cross sectional area   40 %

The calculation of the storage and their performance is based on the following refractory materials in the checker work:

        3,6 m of high alumina bricks, > 62 % Al2O3
                in the upper part of the checker work

        5,4 m of fireclay bricks, > 35 % Al2O3
                in the lower part of the checker work

The lining in the shell of the storage (insulation) is as follows: (from inside the innermost face of the storage)

        high alumina brick, > 62 % Al2O3   230 mm

        insulating refractory bricks   3 x 114 mm
        (different qualities, depending on
          the temperature in the wall)

        insulating bricks   230 mm

        steel 15 MO 3   45 mm

The complete thickness of the wall (with expansion joints) is 860 mm. The external diameter of the steel shell is 3,97 m.

Figure 24 gives the temperature in the storage wall for an internal temperature of 1200 deg C.

The internal diameter of the dome is 2,71 m. This is because the inside layer of the high alumina brick must expand. The layers are the same as in the storage wall. A schematic diagram of the insulation is shown in the layout drawing no. 86087 A.

The whole height of the storage with inlets and without the dome is 12 m. With a temperature of 1200 deg C in the dome and a cold temperature of 250 deg C at the lower end the heat losses of one storage are 34 kW. The heat losses in relationship to the inside wall temperature are shown in figure 25.

5.06   CONNECTING MAINS

5.06.01   RECEIVER / MUSHROOM VALVE NEAR STORAGE

operating temperature   1200 deg C

design temperature   1400 deg C

operating pressure   approx. 1,1 bar

internal diameter   500 mm

lining of the wall (from inside):

     refractory castables   120 mm

     ceramic fibre blanket   230 mm
     (different qualities, depending on
      the temperature in the wall)

     steel   12 mm

Heat losses at operating temperature (assumed at the wall):
1,389 kW/m.

The temperatures in the wall are shown in figure 26 A temperature cycle over a day will be discussed later.

5.06.02   VALVE / STORAGE

operating temperature   1200 deg C

design temperature   1400 deg C

operating pressure   max.   20 bar

internal diamter   500 mm

lining of the wall (from inside):

    refractory castables   160 mm
    (different qualities, depending on
     the temperature in the wall)

    insulating castable   190 mm

    steel   12 mm

Heat losses at operating temperature (assumed at the wall):
3,049 kW/m.

5.06.03   STORAGE / BLOWER / RECEIVER

operating temperature   300 deg C

operating pressure   approx. 1,1 bar

internal diameter   500 mm

internal diameter   500 mm

Lining of the wall (from inside):

     steel   12 mm

     insulation (e.g. fibre blanket)   250 mm

Heat losses at operating temperature: 0,123 kW/m.

5.06.04   STORAGE_/_STACK

operating temperature   300 deg C

operating pressure   approx. 1,1 bar

internal diamter   200 mm

Lining of the wall (from inside):

     steel   5,9 mm

     insulation (contact safety)   50 mm

Heat losses at operating temperature: 0,547 kW/m (not significant).

5.06.05   COLD_BLAST_(PRESSURE)_/_STORAGE

operating temperature   200 deg C

operating pressure   20 bar

internal diamter   100 mm

Lining of the wall (from inside):

    steel   3,6 mm

    insulation (e.g. fibre blanket)   250 mm

Heat losses at operating temperature: 0,062 kW/m.

5.06.06   STORAGE_/_CONSUMER

operating temperature   1000 deg C

design temperature   1400 deg C

operating pressure   20 bar

internal diameter   100 mm

Lining of the wall (from inside):

    SiSiC for inside lining (protection)   4 mm

    insulation fibre blanket   96 mm

    steel   7,5 mm

Heat losses at operating temperature: 0,548 kW/m.

A Sketch of the inside protection liner (SiSiC) is shown in figure_27.

5.06.07   COLD_BLAST_MIXING

operating temperature   200 deg C

operating pressure   20 bar

internal diameter   21 mm

Lining of the wall (from inside):

      steel   2 mm

      insulation (e.g. fibre blanket)   50 mm

Heat losses at operating temperature: 0,068 kW/m.

5.07   DESIGN_OF_THE_COMPONENTS

5.07.01   PRESSURE_UNITS

2 pressure units are needed in the cycle storage / consumer
(20 bar). Firstly, a compressor with a compressedair receiver
to fill a storage with air of 20 bar and secondly, a booster
for the pressure circulation from storage to consumer and back
to storage.

In the calculations for the costs the following components are
are assumed. Some alternatives and the costs are discussed
later.

In addition to the pressure units a silencer is necessary when
a storage changes from unloading (blast) to loading from the
receiver.

5.07.01.01   COMPRESSOR FOR ATMOSPHERIC AIR

flow rate   822 Nm3/h

pressure   20 bar abs.

It is to be assumed that the oil content of the air after
compression will be a max. of 8 mg/Nm3.

5.07.01.02   COMPRESSEDAIR RECEIVER

volume   5 m3

temperature after receiver for
    compressedair   max. 200 deg C

A bypass around the compressedair receiver is necessary so as
to achieve a short period for filling a storage.

The time to fill a storage is nearly 16 minutes. There are
several possible ways of achieving a shorter filling time:

    - a larger compressedair receiver,

    - a compressor with a higher flow rate,

    - a compressor and a compressedair receiver designed
      for a higher pressure.

The costs for a compressedair receiver with an additional 5 m3
are approx. 9000 DM. The tine will decrease by approx. 3 or 2
minutes (depending on the volume of the compressedair
receiver).

To increase the flow rate of compressed air by about 50 % the
costs of the compressor will increase by about 50000 DM. The
time will decrease by approx. 4 minutes. The total energy to
compress the air for a storage is the same in both these two
cases.

To increase the pressure in the compressedair receiver and at
the outlet of the compressor to fill the storage only from the
receiver itself requires a very great increase in pressure.
For example, with a 15 m3 compressedair receiver the design
pressure for the compresser must be 40 bar (30 m3 needs 30
bar). The costs will increase extremly.

It would be preferable to have all pressure components
(compressor, oil separator, cooler etc.) together in a
container. The costs for a container would be about 50000 DM.
They are not included in the costs of our calculation.

5.07.01.03   BOOSTER

The booster operates in a closed circuit , storage -->
consumer --> booster --> storage.

medium	atmospheric air
flow rate	800 Nm3/h
operating temperature	200 deg C
design temperature	400 deg C
operating pressure	20 bar
pressure difference	10 mbar

5.07.01.04   SILENCER

When a storage is changed over from unloading or discharging
(which occurs several times a day) the pressure must be
discharged. Calculations have shown that discharging into a
compressedair receiver is not recommendable because the
pressure in the compressedair receiver is mostly above 1 bar.

It will be about 10 bar (depending on the size of the compressedair receiver).

The silencer is designed as follows: (see figure 28)

pressure in storage   20 bar

medium atmospheric air

operating temperature   200 deg C

design temperature   400 deg C

duration of pressure release approx.   8 min.

noise level at 50 m   55 dB(A)

5.07.02   CONSUMER

In the first instance a special consumer or user of the high quality energy stored in the three storages is not planned. A turbine for these flow rates is not available. The pilot storage plant in Almeria will show that a process with 1000 deg C and 20 bar can work with solar energy with storages.

In the pressure cycle a heat exchanger is necessary to cool the hot air from 1000 deg C to 200 deg C. In a simplest way the hot air (1000 deg C) is first mixed with cold air to 800 deg C and it is then cooled down. The heat exchanger works with thermal oil. The thermal oil is cooled with a simple air cooler. If it should be necessary or desirable, the air cooler can be replaced by another component in order not to destroy the energy but to sensibly use it in the plataforma solar.

For example see figure 29.

5.07.03　　　　BLOWER (pressureless cycle)

Medium, atmospheric air

flow rate	max.	2500 Nm3/h
operating temperature		300 deg C
design temperature		400 deg C
pressure difference		100 mbar
design pressure		1 bar abs.

The blower is needed to transport the air through the receiver, the connecting mains, the storage or storages, the cold connecting mains to the blower itself and then back to the the receiver. A pressure drop of 50 mbar is assumed for the receiver (after final design of the receiver the pressure difference for the blower must recalculated).

5.07.04　　　　HEATING SYSTEM

A heating system is needed for initially heating up the storages and connecting mains or after cooling down. The heating system can work with an oil or an gas burner.

An oil burner (light oil) is assumed in our calculations. The oil burner fires into a combustion chamber with a refractory lining. An oil container is necessary. The heating system is also needed if the performance of the receiver is not high enough for the consumer. If the tests of the storage plant stop for a short period the heating system can be used to keep the whole storage plant at operating temperature.

The components of the heating system are designed as follows:

performance	approx. 400 kW
medium	light oil
control range	1:3
pressure in combustion chamber	20 mbar over normal
temperature of the waste gases	1200 deg C
operating of the heating system from the control room is normal	
volume of oil container	10000 l

One charge of oil in the oil container (10000 l) can achieve the following alternatives:

- heating up of the storage plant to normal operating temperature,

- maintaining the storage plant at operating temperature for approx. three weeks, when receiver and consumer are not working,

- working with the storage plant with consumer and without the receiver for about one week.

The calculations and results of the heat and energy losses are given in a later chapter of this study.

5.07.05   STACK

When the heating system (oil burner) is operating a stack for the waste gases is necessary. It is designed as follows:

height	25 m
external diameter	700 mm
internal lining, refractory material	

5.07.06   VALVES

All valves are electrically operated. Valves having a design
or operating temperature above 600 deg C are water cooled.

3	mushroom-valves,	DN 1200, PN 20
	internal diameter	500 mm
	design temperatur	1400 deg C
	operating temperature	1200 deg C
3	mushroom-valves,	DN 300, PN 20
	internal diameter	100 mm
	design temperature	1400 deg C
	operating temperature	1200 deg C
3	waste gas shut-off valve	DN 500, PN 20
	external diameter	1000 mm
	design temperature	400 deg C
	operating temperature	300 deg C
3	waste gas shut-off valve	DN 200, PN 20
	external diameter	320 mm
	design temperature	400 deg C
	operating temperature	300 deg C
3	cold blast shut-off valve	DN 100, PN 20
	external diameter	500 mm
	design temperature	400 deg C
	operating temperature	200 deg C
3	filling valve for cold blast	DN 50, PN 20
	design temperature	400 deg C
	operating temperature	200 deg C
3	blow-off valve,	DN 50, PN 20
	design temperature	400 deg C
	operating temperature	200 deg C
2	mushroom valves,	DN 1200, PN 2.5
	internal diamter	500 mm
	design temperature	1400 deg C
	operating temperature	1200 deg C
1	valve for mixing blast	DN 25, PN 20
	design temperature	400 deg C
	operating temperature	200 deg C

1	control valve for mixing blast DN 25, PN 20	
	design temperature	400 deg C
	operating temperature	200 deg C

An example of a mushroom valve is shown in figure 30.

5.08   ENERGY CONSUMPTIONS

5.08.01   ELECTRICAL POWER REQUIRED

The electrical power required for components described in the foregoing chapters is as follows: (approx. values)

compressor (822 Nm3/h)	130 kW
alternative	
compressor (1260 Nm3/h)	196 kW
booster	1 kW
blower	25 kW
heating system (blower for burner)	2,2 kW
valves or mushroom valves	
DN 500 and above	1,50 kW/valve
smaller than DN 500	0,75 kW/valve

It is to be noted that only one valve operates at any given time.

5.08.02   WATER CONSUMPTION FOR VALVES

The flow of cooling water needed for the valves is as follows:

mushroom valve DN 500/1200	36 m3/h per valve
mushroom valve DN 100/300	10 m3/h per valve

There are 5 mushroom valves with DN 500/1200 and 3 mushroom valves with DN 100/300 which are cooled. The total flow of cooling water is about 210 m3/h.

5.08.03   ELECTRICAL ENERGY CONSUMPTION

The energy consumption is given for one day. It is assumed that the storages are changed from 1 to 20 bar 9 times a day (3 times per storage).

compressor (822 or 1260 Nm3/h)	337 kWh/day
booster (pressure cycle)	1 kWh/day
blower (receiver cycle) 21.06.	291 kWh/day
21.12.	194 kWh/day

The electrical energy consumption for the valves is not significant.

The energy consumption of the heating system is dependent on the strategy of operating the storage plant.

Other electrical energy such as lights etc. are not significant, but the installation of these components is necessary.

5.08.04   HEAT LOSSES

The total heat losses (3 storages with 34 kW per storage and the connencting mains) is calculated at 200 KW.

The maximum power of the receiver (PR in chapter 5.04) must be designed as follows depending on the date:

date	tD	average over time tD	max (PR)
21.06.	12	1000 kW	1375 kW
21.03.	10	1200 kW	1649 kW
21.12.	8	1500 kW	2061 kW

5.08.05   TEMPERATURES IN THE STORAGE

For example the the temperatures in the pilot storage (checker work) are given for a performance of a day 21.03. with a loading / unloading cycle similar to figure 23.

Figure 31 gives the temperatures in the storage when it is loaded in 5 hours.

Figure 32 shows the temperatures when the storage unloads over 8 hours.

Figure 33 shows the temperature at the storage outlet. It is assumed that the storage is loaded at 1200 deg C.

5.08.06   TEMPERATURES IN A CONNECTING MAIN

The connecting main (pipe) from receiver to mushroom valve has
been taken as an example.

In figure 34 the temperatures in the wall of the main are
given over several days. The wall has 3 different layers.

Figure 35 shows the temperatures at the inside surface and of
the air over 12 h of blowing for some points of the main,
namely, at the inlet, after 1/3 of length, after 2/3 of length
and at the outlet.

In figure 36 the wall and air temperatures are given over the
length of the main for several hours.

Figure 37 shows how the wall of the main cools down when the
plant or the receiver stops over a period of several days.

5.08.07   HEATING UP THE STORAGES

5.08.07.01   INITIAL HEATING UP

To heat up the 3 storages and the connecting mains (heat
losses and waste gas assumed) the energy consumption is about
100 MWh. The connecting mains and storages have to be
initially dried for one week (up to about 100 deg C). After
that the components must be heated up to operating temperature
in two weeks (linear from 100 deg C to 1200 deg C at top of
the storages).

5.08.07.02   SUBSEQUENT HEATING UP

If the storages are dry the heating up procedure needs two weeks for heating up to operating temperature. The energy consumption is less but still about 100 MWh.

It is now to be calculated whether the energy situation is better if the storages are cooled down and afterwards reheated when the whole plant is stopped for more than 3 weeks.

It must be noted that a cooling down and heating up procedure is not a normal operation for storages with refractories. Consideration must therefore be given to the technical risk of the cooling down procedure. Perhaps it is better to keep the storage on operating or near operating temperature over 4 or 6 weeks.

5.09   OPERATION IN CASE OF DISTURBANCES

Figures 38, 39, 40 and 41 show a list what to do in case of disturbances:

figure 38	storage
figure 39	receiver
figure 40	consumer
figure 41	general

5.10   ESTIMATED COSTS (PILOT STORAGE PLANT)

The estimated costs for   the   pilot storage plant   in Almeria
are:

     17 216 000 DM.

The following items are not included in the price:

     foundation

     anchoring of the storages

     electric circuit, pipes for water, etc.

     heating media (oil, electricity, etc.)

     lights for the plant

     telephone

The costs are estimated costs of today.

6   COMMERCIAL HEAT STORAGE

6.01   DESIGN OF ONE COMMERCIAL HEAT STORAGE

The commercial heat storage has been designed as follows:

- design pressure, 10 bar,

- the height of the storages, about 40 m.

It is possible to put the receiver on top of the storage with the following advantages:

- the connecting mains from the receiver to the storage are very short,

- with short connecting mains the heat losses are very low,

- the performance of one receiver is only a part of the performance of the plant, because one receiver is needed for each storage.

A sketch of one storage is given in figure 42. The height of the receiver in relationship to the height of the storage can be changed. The alternatives in this sketch are as follows:

- the receiver is on top of the storage (above the dome).

- the receiver is adjacent to the storage at the upper end of the checker work, this alternative is more expensive than the other (higher).

For 10 bar pressure it is possible to have a normal water cooled hot blast valve or a mushroom valve. The first alternative (on top of dome) needs a normal hot blast valve because the pressure is inside the storage and therefore with a mushroom valve the design of the connecting main is very difficult. For the other alternative both valves are possible but a normal hot blast valve is not so difficult.

A sketch of a normal hot blast valve (water cooled) is given in figure 43.

The checker work and the storage is similar to the checker chamber in a hot blast stove plant built by DIDIER in 1985 (ROGESA, Dillingen).

The checker work for the design case is calculated as follows (see also pilot storage plant) :

     weight of the checkers   approx. 2280 t

     height of the checker work   35,4 m

     diameter of the checker work   7,7 m

     free cross sectional area   40 %

The calculation of the storages and their performance is based on alumina and fireclay bricks.

The lining in the shell of the storage (insulation) is similar to that of the pilot storage plant: (from the inner side of the storage)

     high alumina brick, > 62 % Al2O3   230 mm

     insulating refractory bricks   3 x 114 mm
     (different qualities, depending on
      the temperature in the wall)

     insulating bricks   230 mm

     steel 15 MO 3   50 mm

The complete thickness of the wall (with expansion joints) is 865 mm. The external diameter of the steel shell is 9,43 m.

6.02   DESIGN OF A COMMERCIAL HEAT STORAGE PLANT

In the following text the commercial storage plant is abbreviated to C S P.

The minimum number of storages in a plant is three (see also under pilot storage plant).

The number of storages could be more than three. The field of heliostats must be optimized if a larger commercial plant with a lot of storages is built up.

6.03

The operation of the storages themselves is similar to that of
the storages in the pilot plant. Figure 44 shows another
cycle. Each storage unloads for 8 hours a day. The energy from
the heliostats (sun) will be divided onto the receivers on top
of the storages. The figure gives the part of the maximum
power of the heliostat field for a day (21.12.) with 8 hours
of sunshine.

The time when a storage is unloading is indicated.

When a storage must change from loading to unloading the
heliostats must be driven from one to an other receiver on top
of the storages.

6.04   TEMPERATURES IN THE STORAGE

For example, the temperatures in the pilot storage (checker
work) are given for a performance of a day 21.03. with a
loading / unloading cycle similar to figure 23.

In figure 45 the temperatures in the storage are given when
the storage is loaded in 5 hours.

Figure 46 shows the temperatures when the storage is unloading
over 8 hours.

Figure 47 shows the temperature at the storage outlet. It is
assumed that the storage is loaded at 1200 deg C.

With the given 3 storages in chapter 6.01 the following
performance is possible (pressure 10 bar):

case I :

temperature in storage from receiver 1200 deg C

temperature to consumer   1000 deg C

temperature to storage lower end   200 deg C

case II:

temperature in storage from receiver 800 deg C

temperature to consumer   600 deg C

temperature to storage lower end   200 deg C

The energy difference in a storage between load and unload is:

I : 77,3 MWh

II : 70,8 MWh

performance consumer:

I : (reference temperature 50 deg C) 10,88 MW

(reference temperature 200 deg C) 9,66 MW

II : (reference temperature 50 deg C) 11.13 MW

(reference temperature 200 deg C) 8,85 MW

The following maximum performance of the whole heliostat field is needed (approx.):

I : reference temperature 50 deg C

    21.12. (8 hours sun)   50 MW

    21.03. (10 hours sun)   40 MW

    21.06. (12 hours sun)   33 MW

I : reference temperature 200 deg C

    21.12. (8 hours sun)   44 MW

    21.03. (10 hours sun)   35 MW

    21.06. (12 hours sun)   29 MW

II : reference temperature 50 deg C

    21.12. (8 hours sun)   51 MW

    21.03. (10 hours sun)   41 MW

    21.06. (12 hours sun)   34 MW

II : reference temperature 200 deg C

    21.12. (8 hours sun)   40 MW

    21.03. (10 hours sun)   32 MW

    21.06. (12 hours sun)   27 MW

The estimated costs for a commercial storage plant with 3
storages are:

    79 000 000 DM

The following items are not included in the price:

    consumer and booster

    foundation

    anchoring of the storages

    electric circut, pipes for water, etc.

    heating media (oil, electricity, etc.)

    lights for the plant

    telephone

The costs are first estimated costs of today.

7   FIGURES

HOT BLAST STOVE WITH INTERNAL COMBUSTION CHAMBER	fig. 1

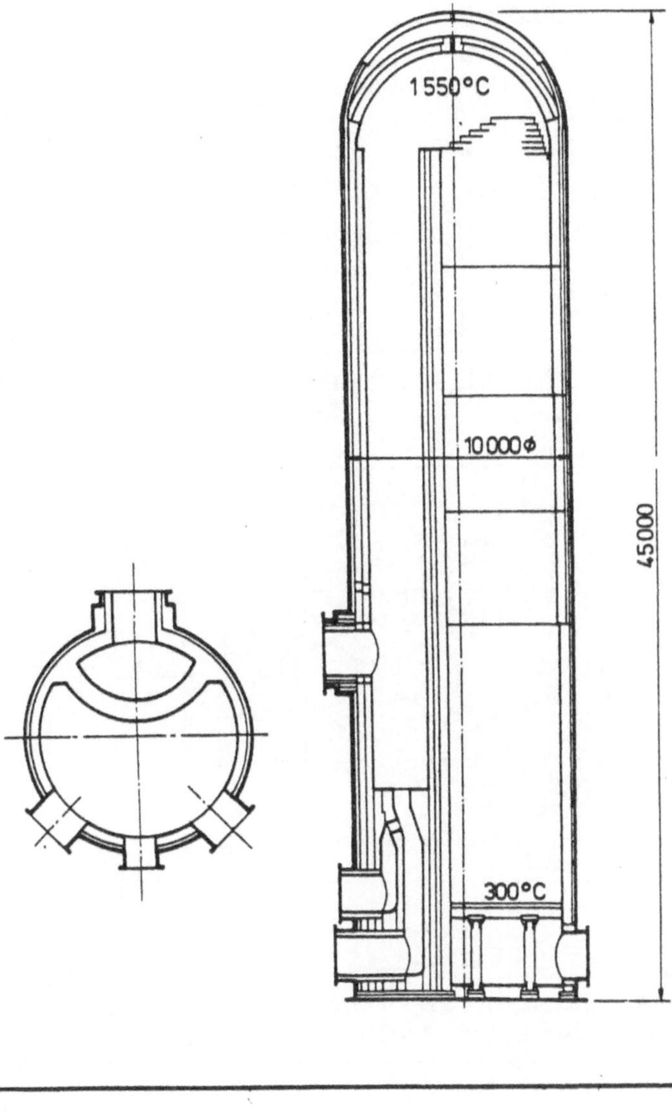

1550°C
10000⌀
45000
300°C

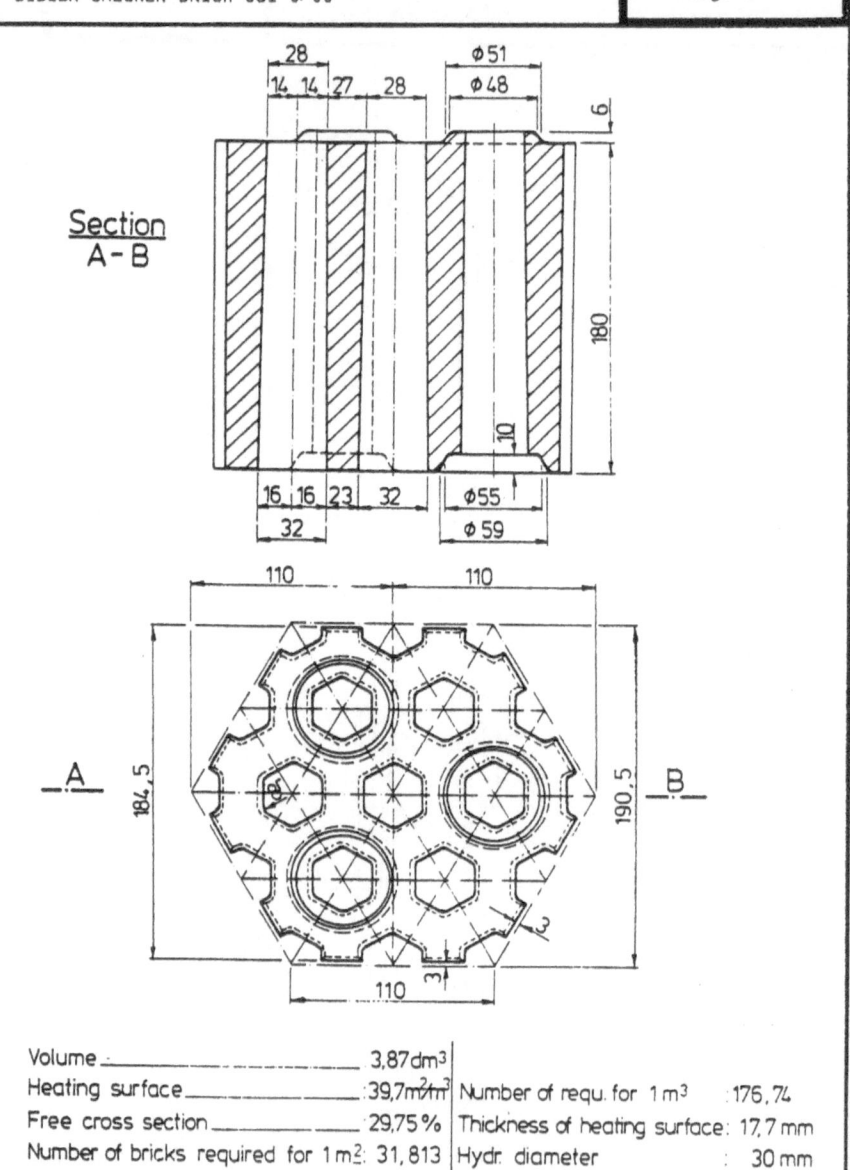

Section
A-B

Volume _____ 3,87 dm³
Heating surface _____ 39,7 m²/m³ | Number of requ. for 1 m³ :176,74
Free cross section _____ 29,75% | Thickness of heating surface: 17,7 mm
Number of bricks required for 1 m²: 31,813 | Hydr. diameter : 30 mm

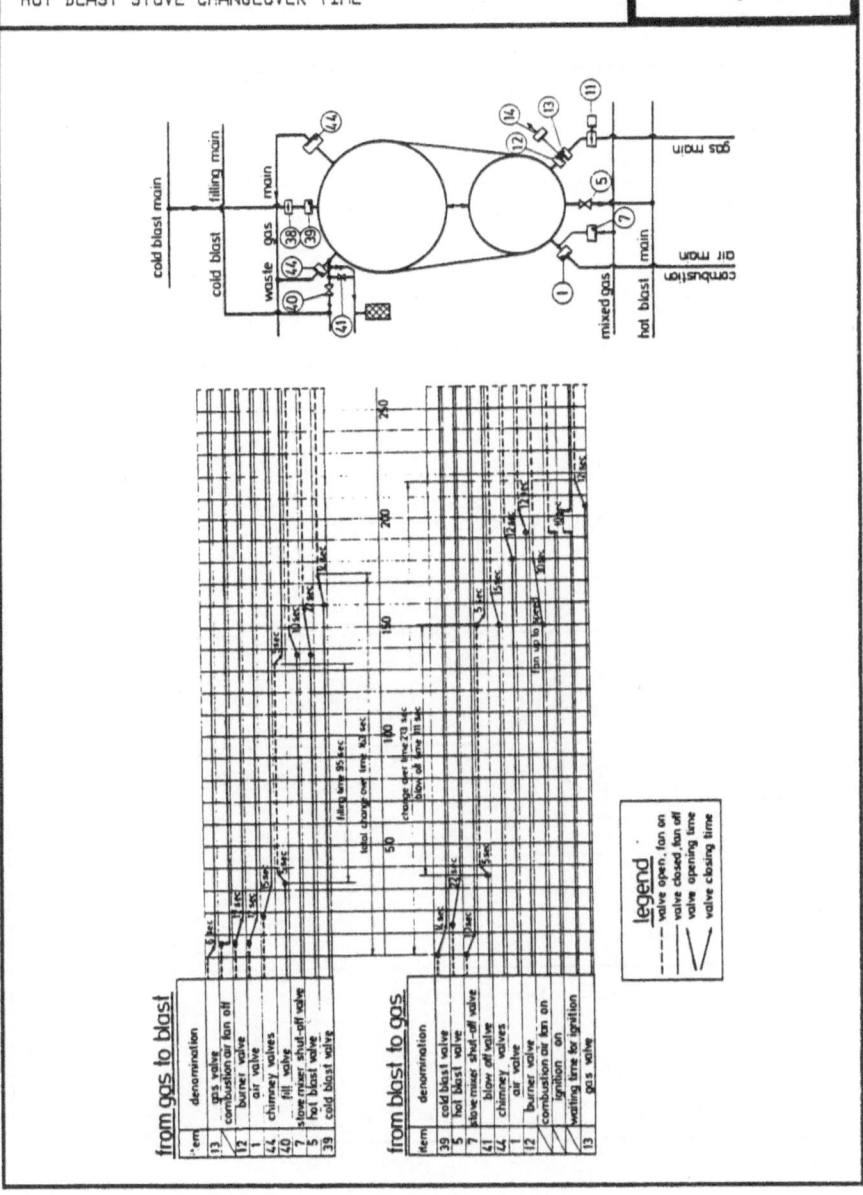

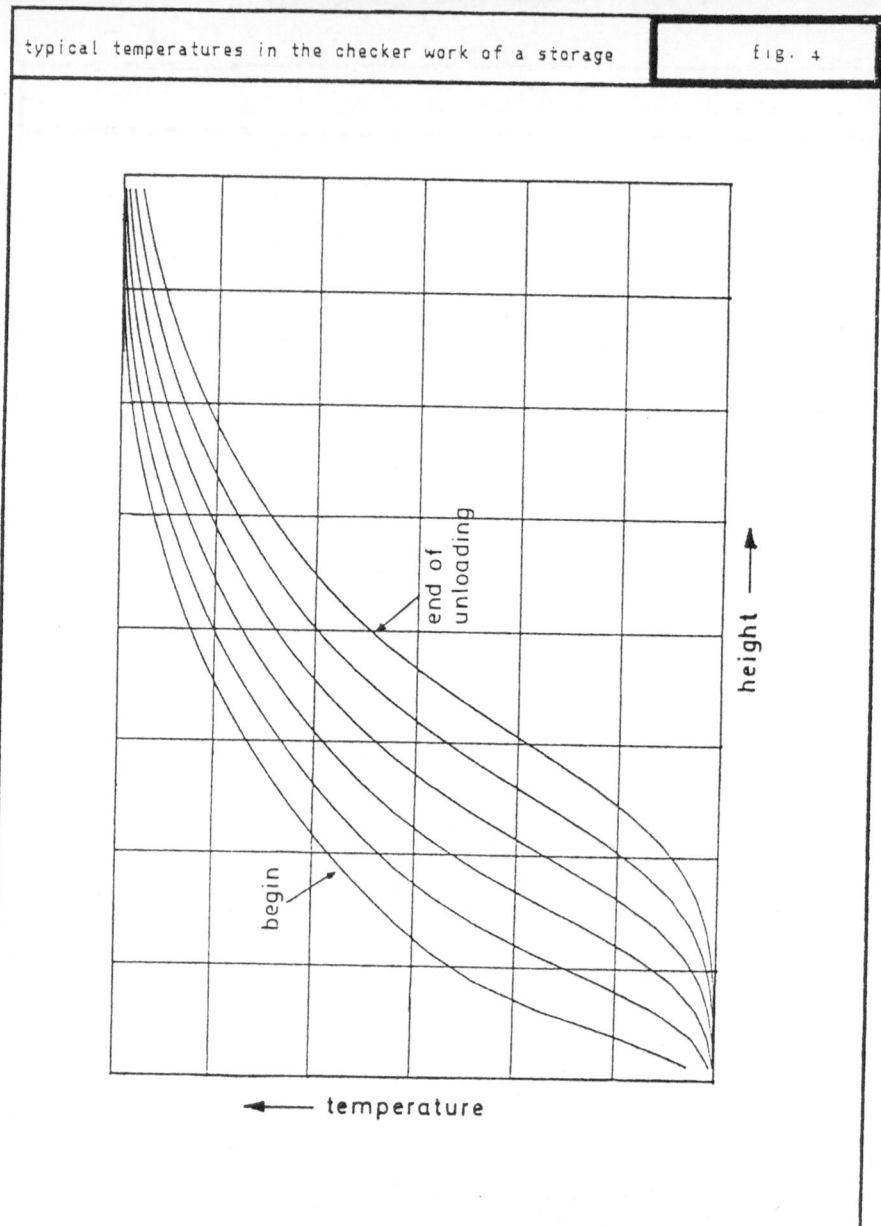

fig. 5

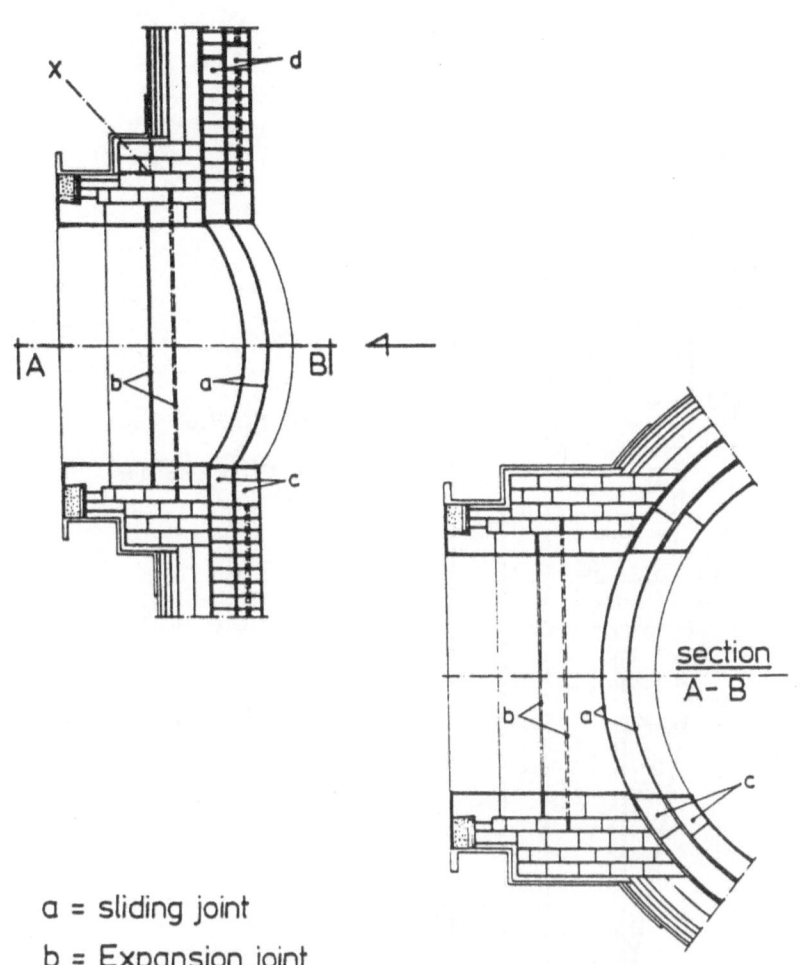

a = sliding joint

b = Expansion joint

c = Special shaped ring

d = Relieving arch

fig. 6

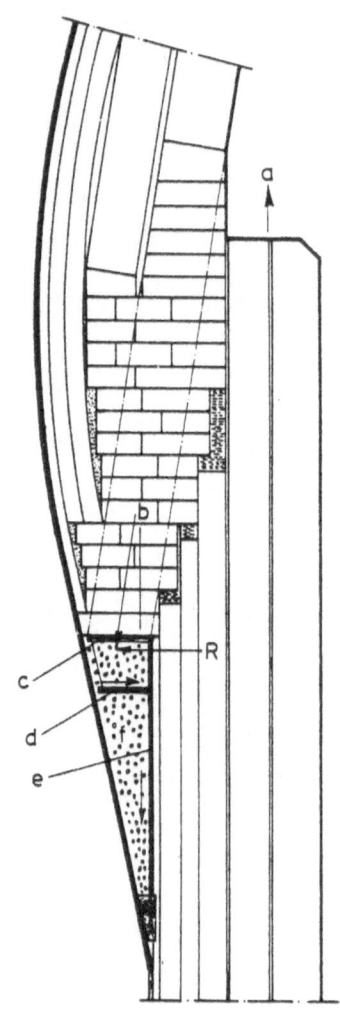

a = Free space

b = Line of pressure

c = Console

R = Resultant action

c-d = Double ring

e = Inner shell

f = Ramming mass filling

fig. 7

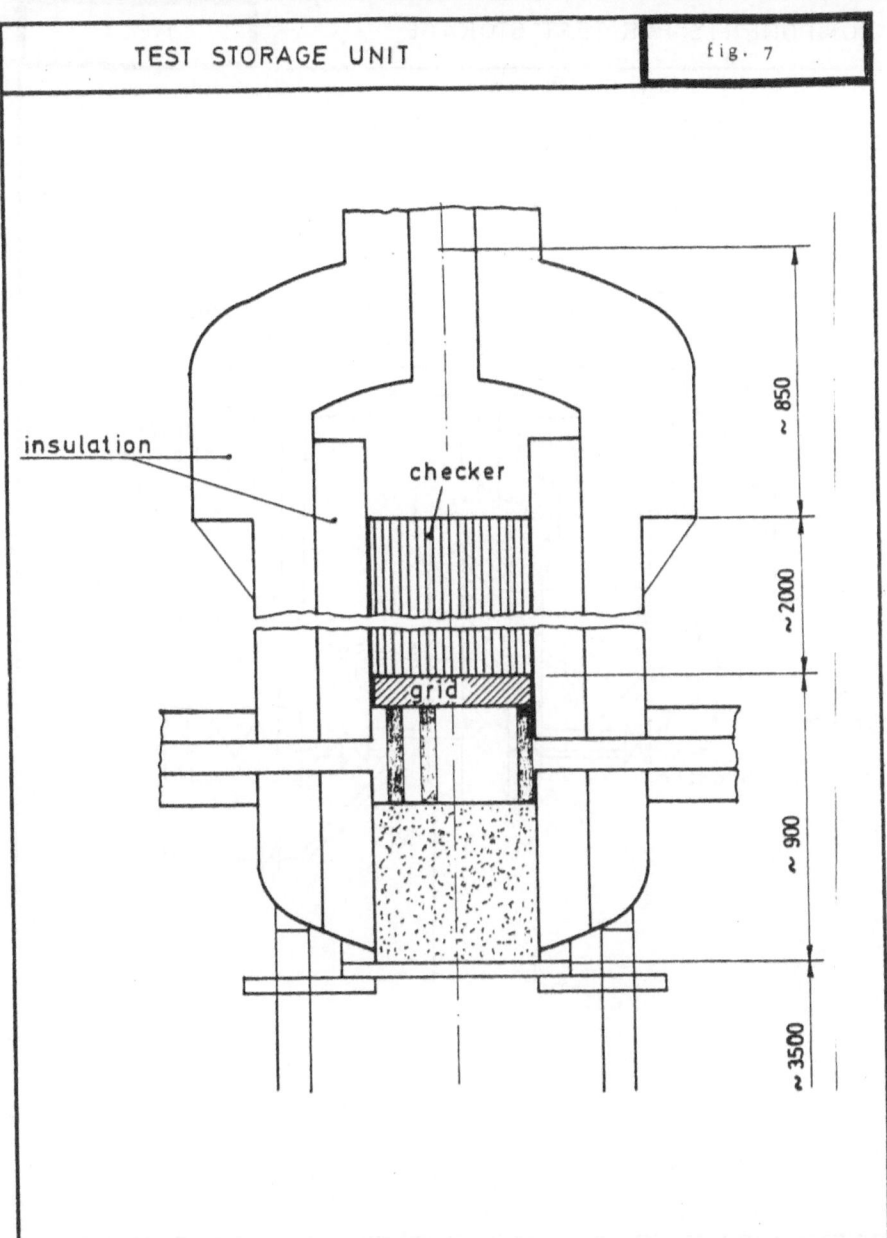

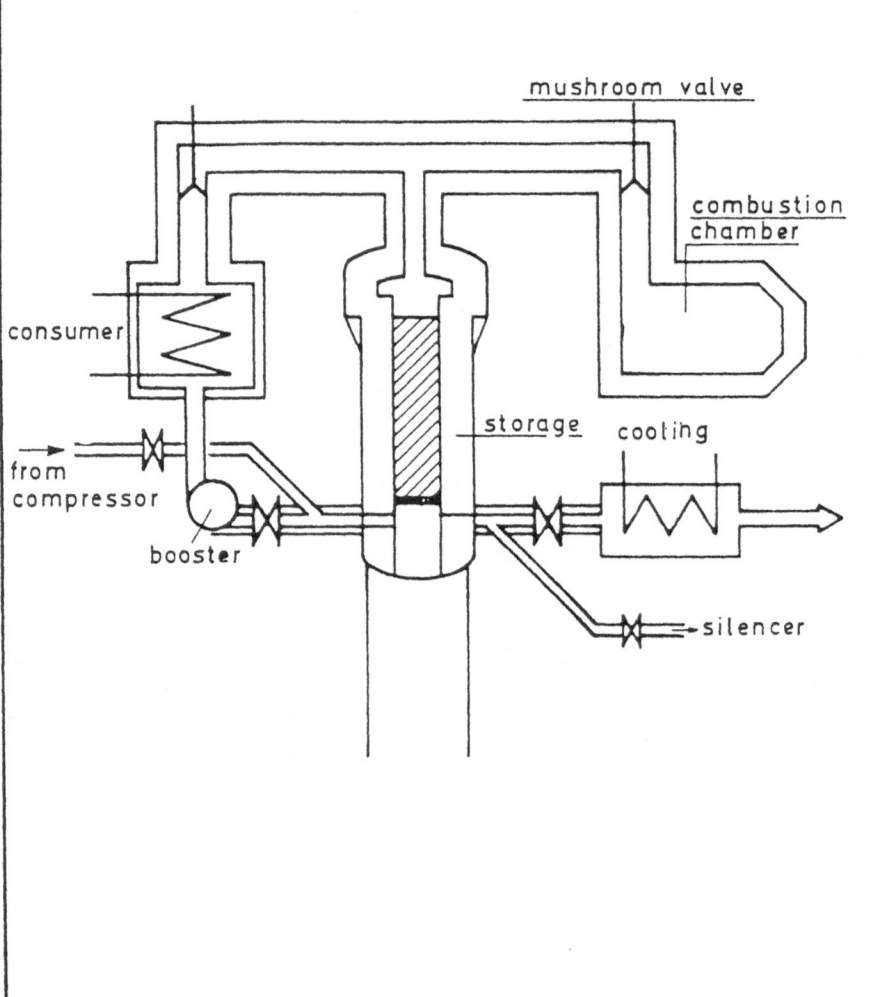

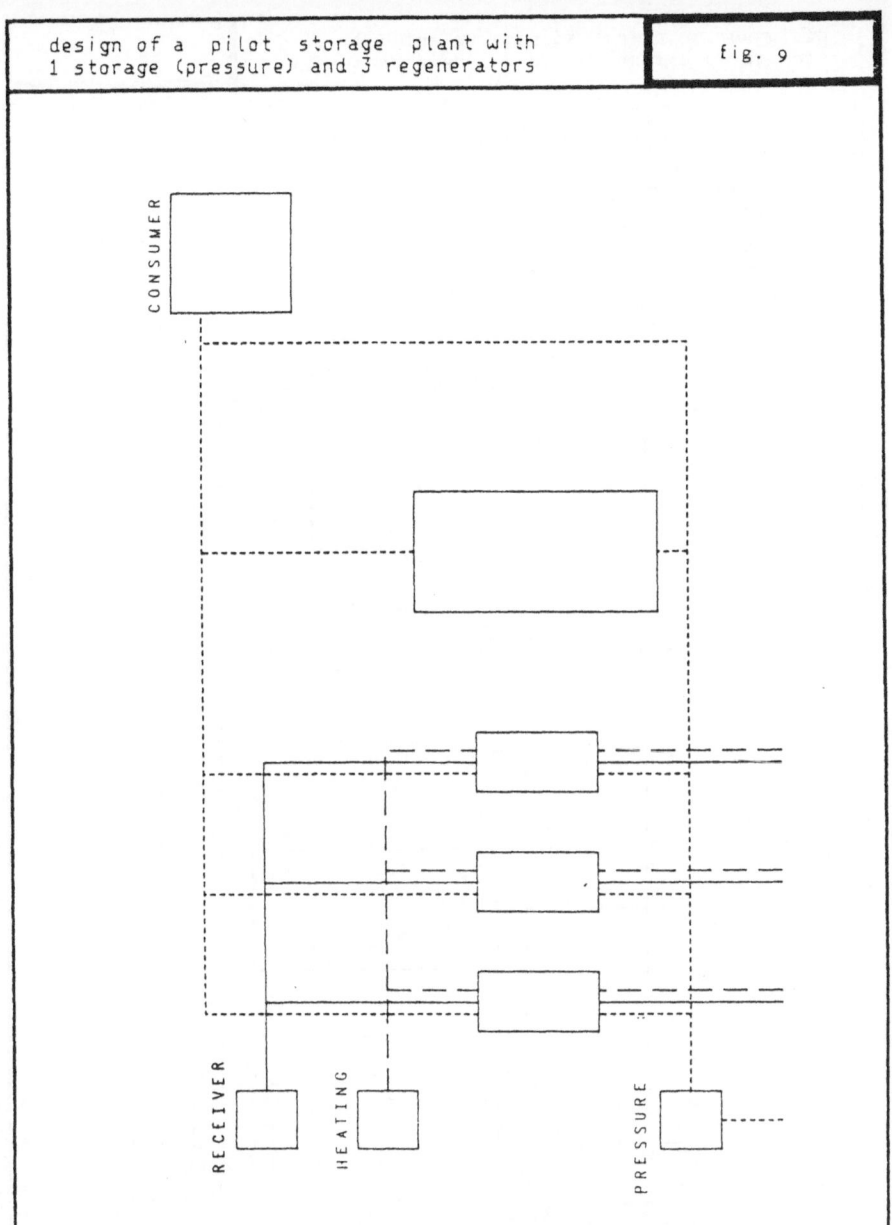

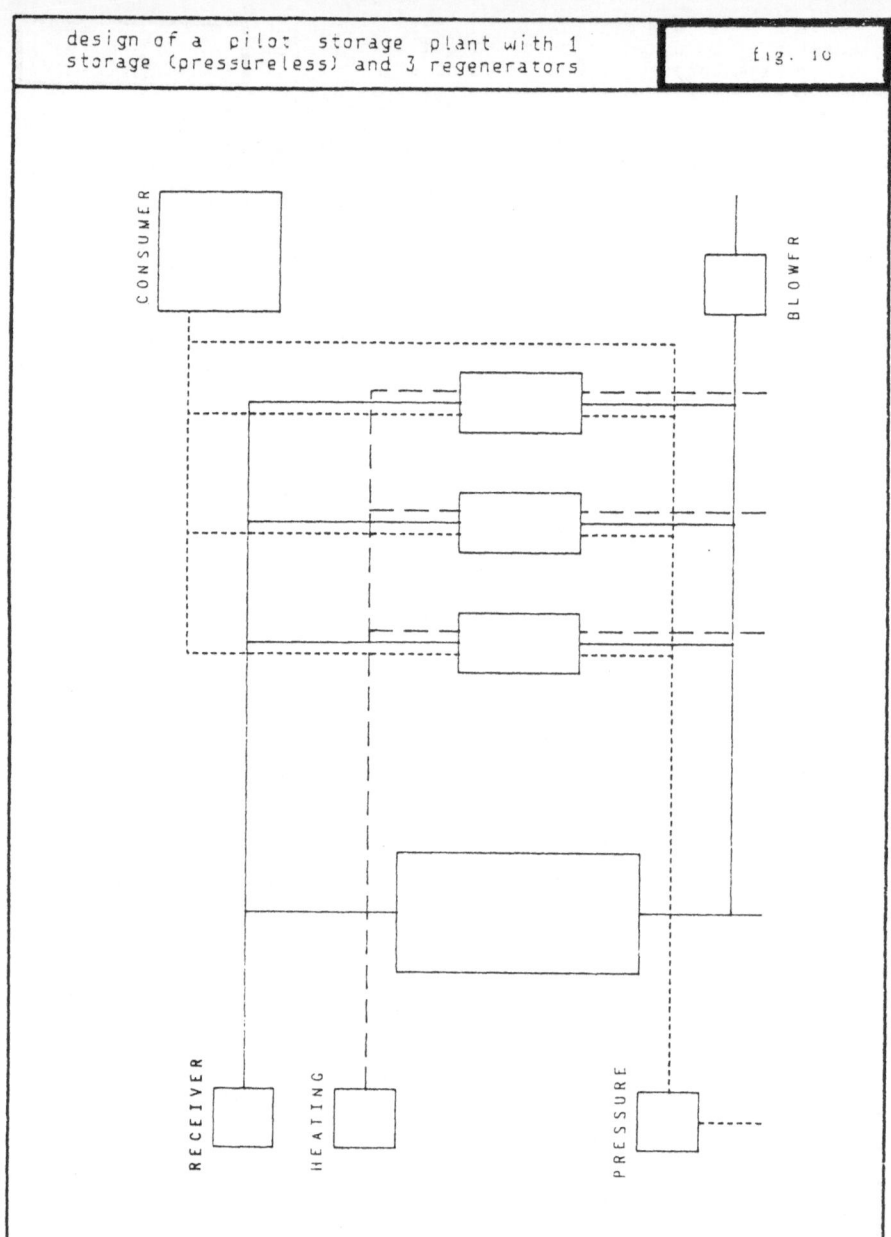

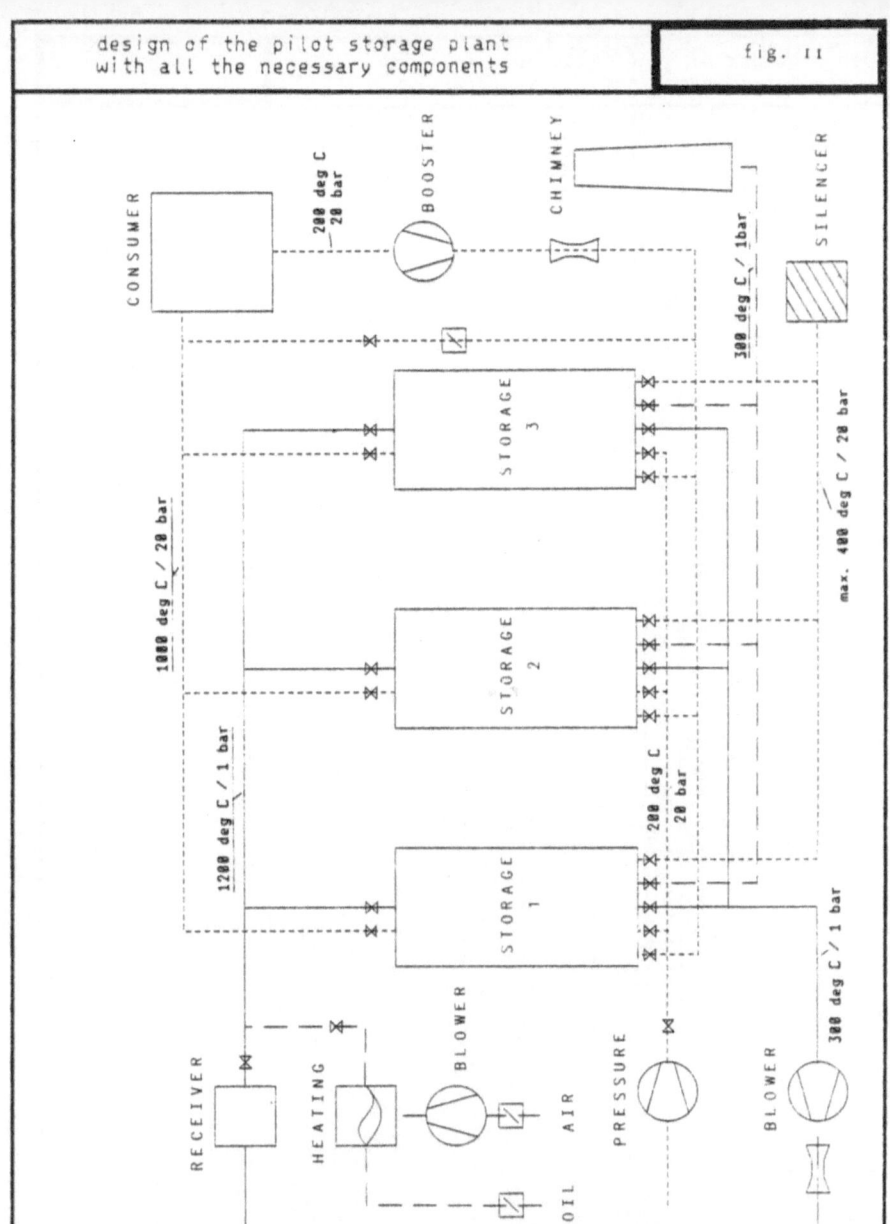

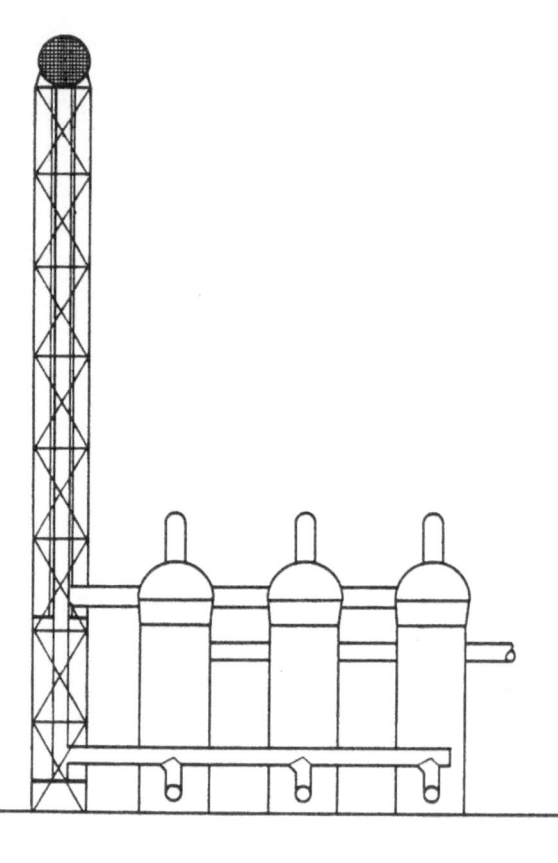

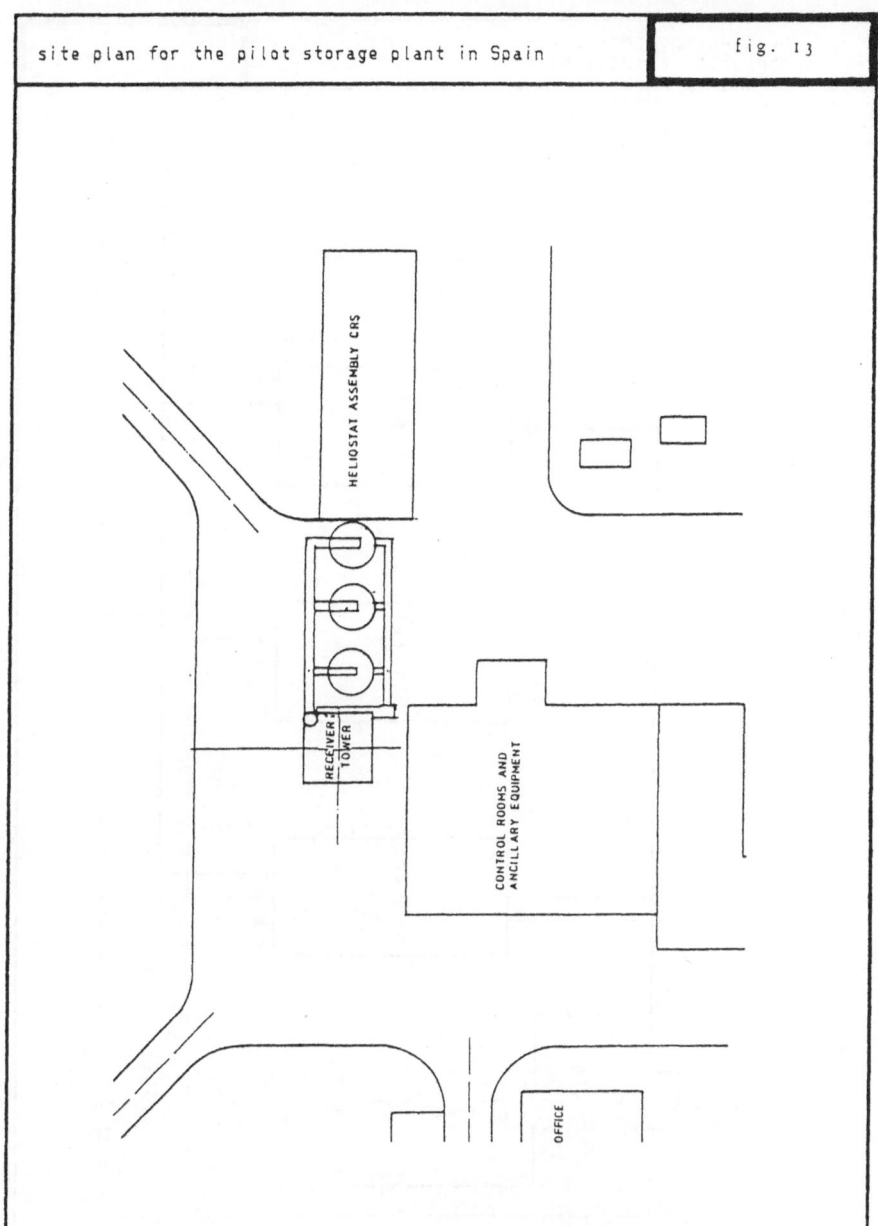

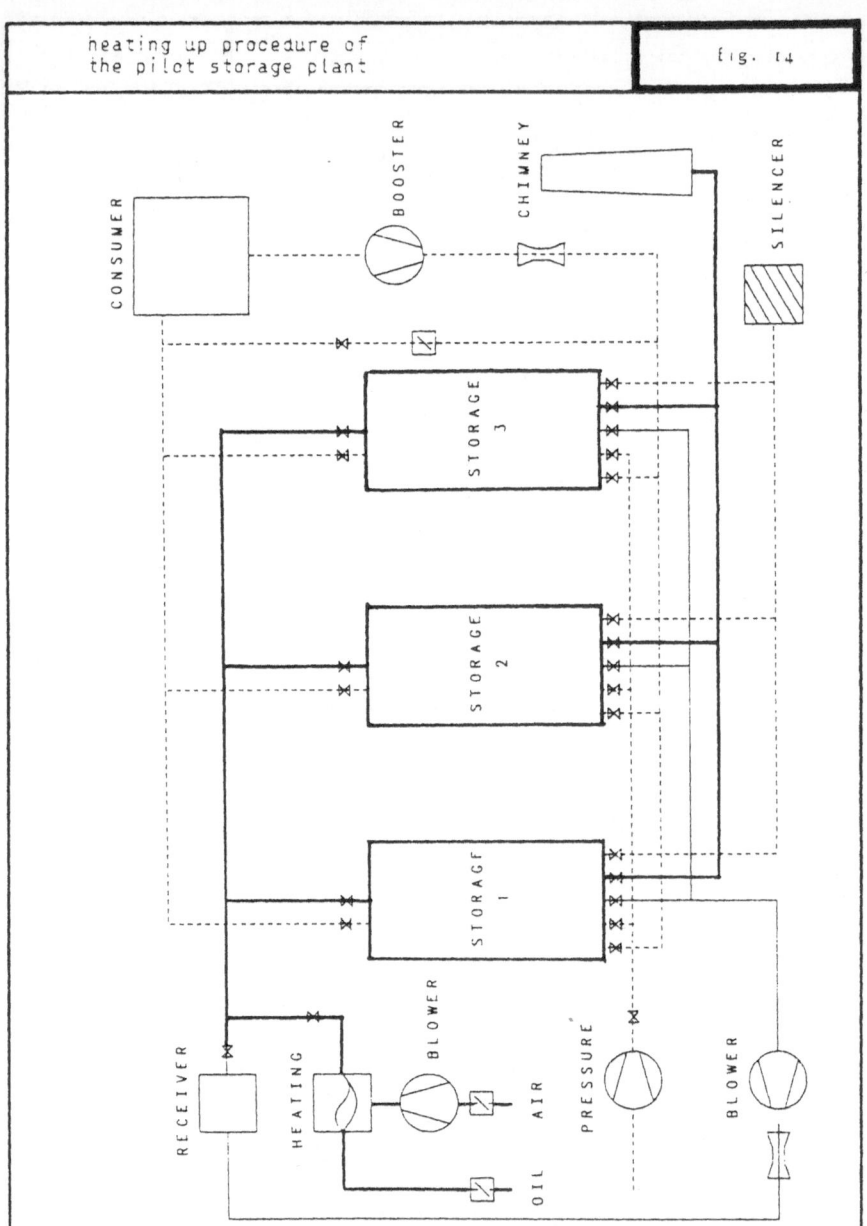

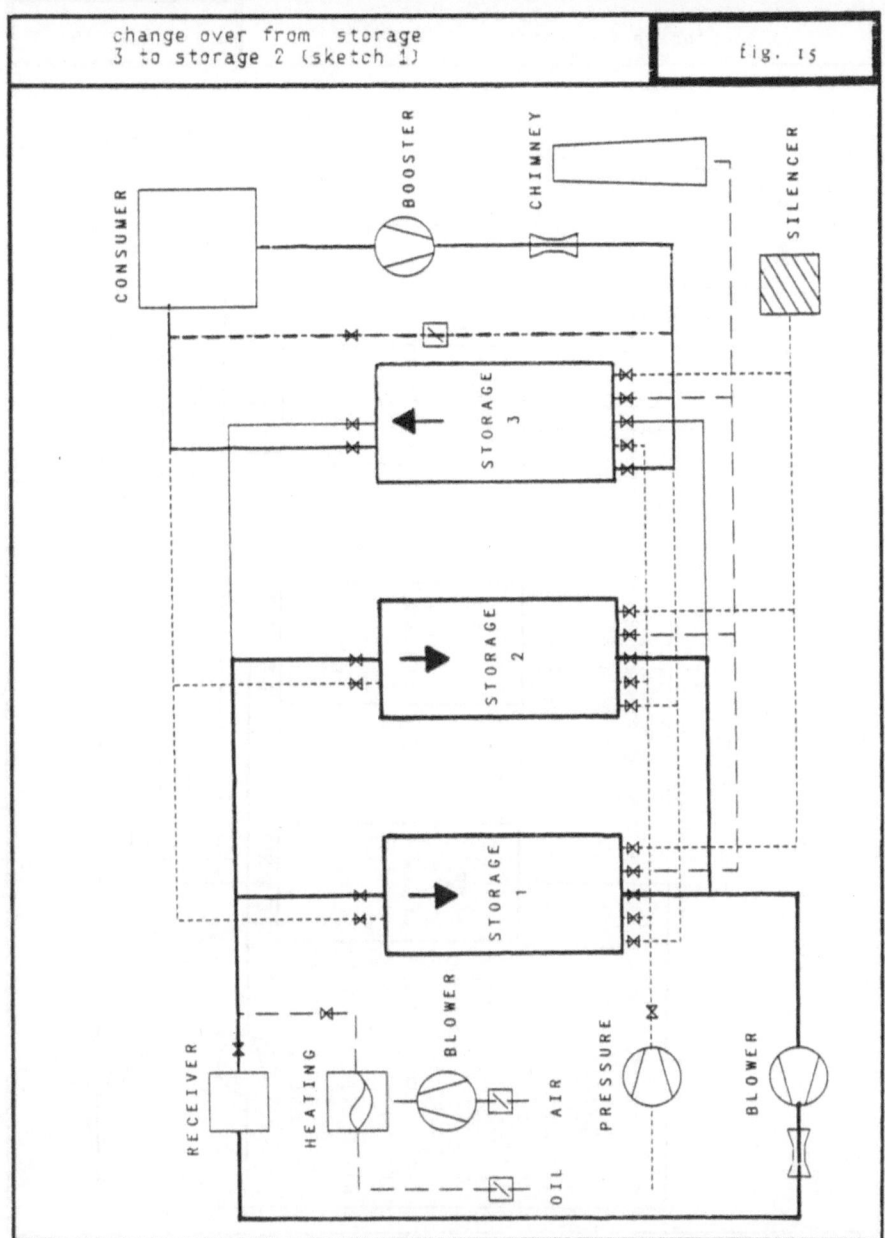

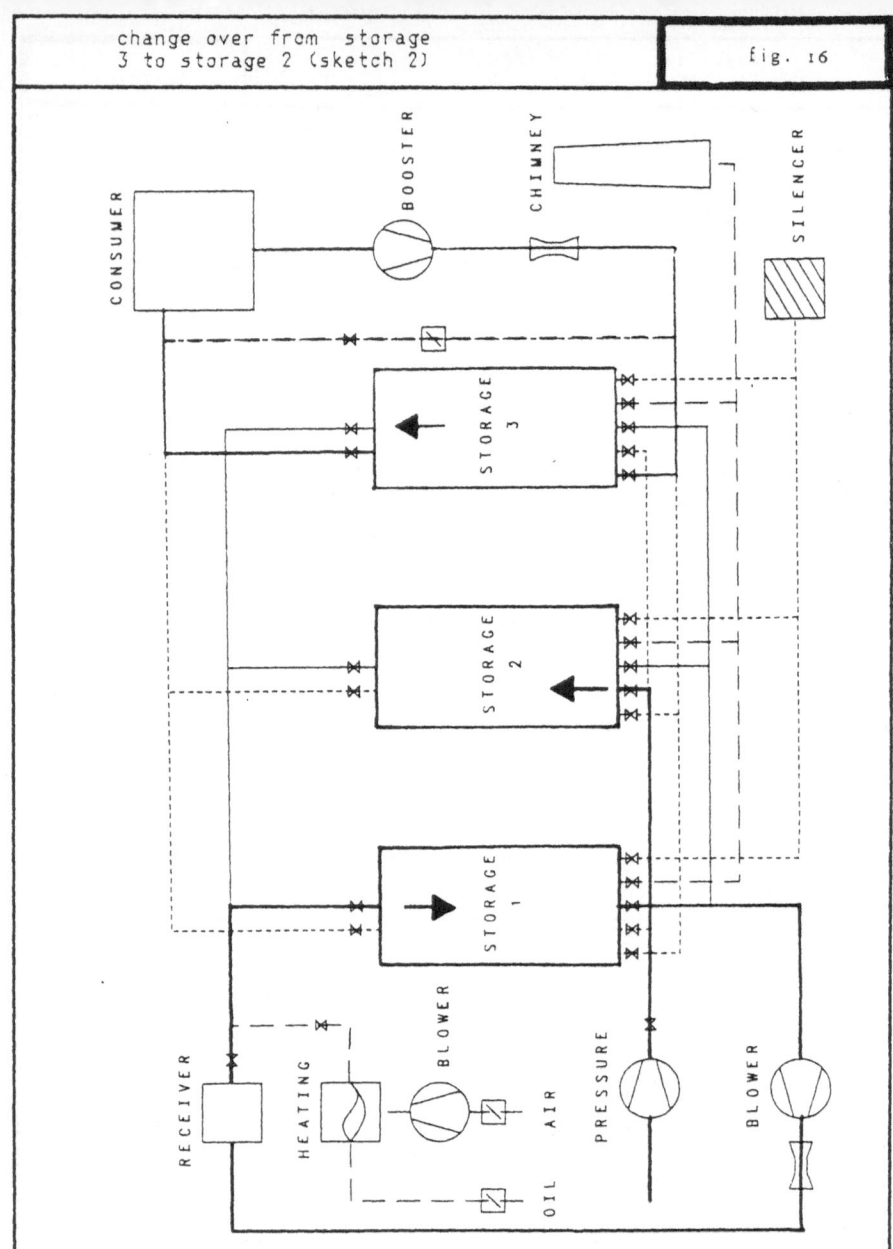

change over from storage 3 to storage 2 (sketch 2)

fig. 16

- 168 -

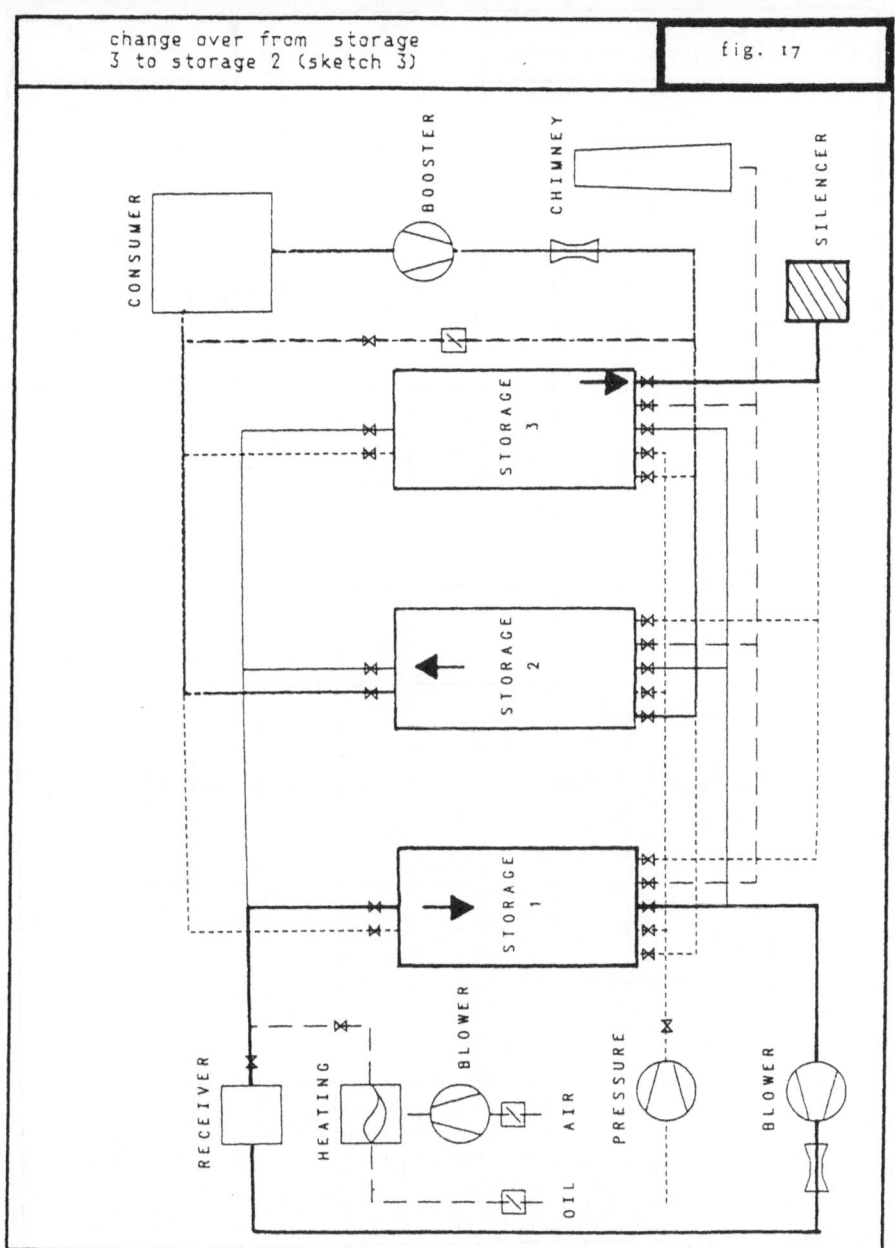

change over from storage
3 to storage 2 (sketch 3)

fig. 17

CONSUMER

BOOSTER

CHIMNEY

SILENCER

STORAGE 3

STORAGE 2

STORAGE 1

RECEIVER

HEATING

BLOWER

AIR

OIL

PRESSURE

BLOWER

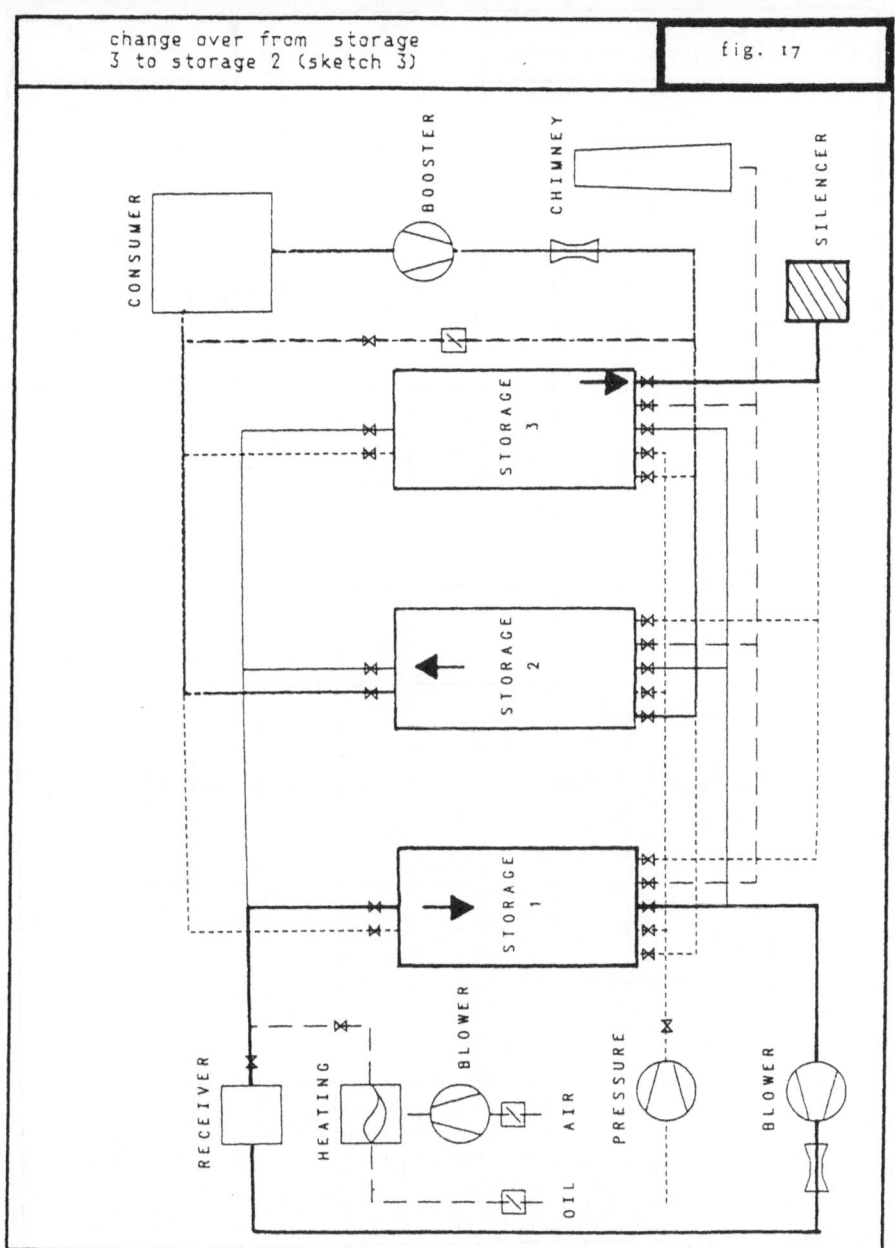

pressure in storage while changing from unload to load

fig. 18

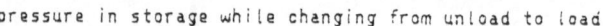

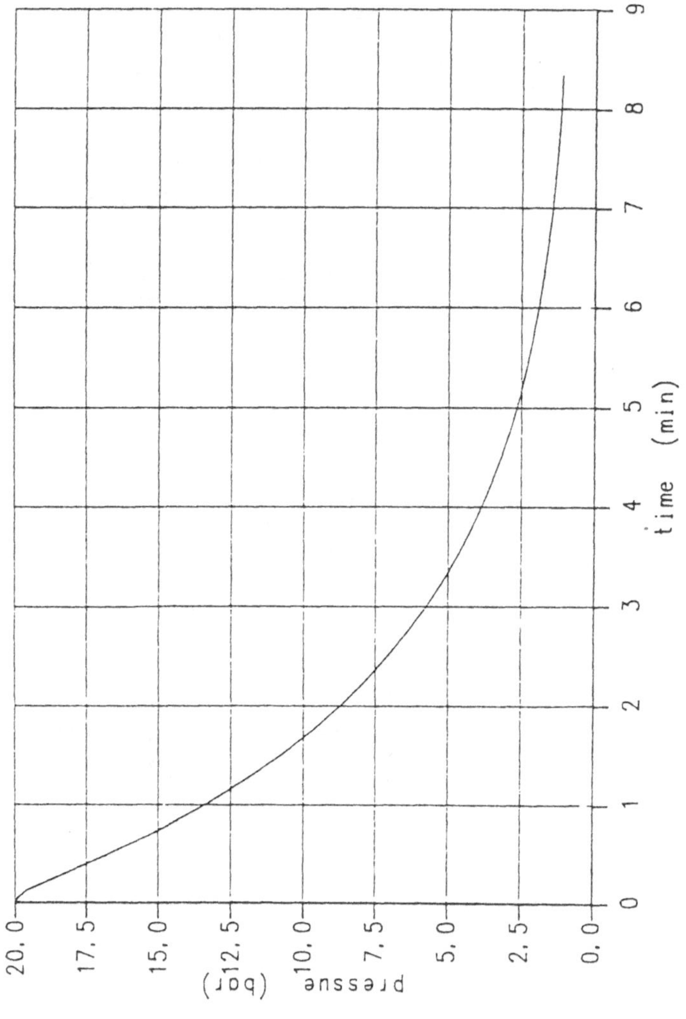

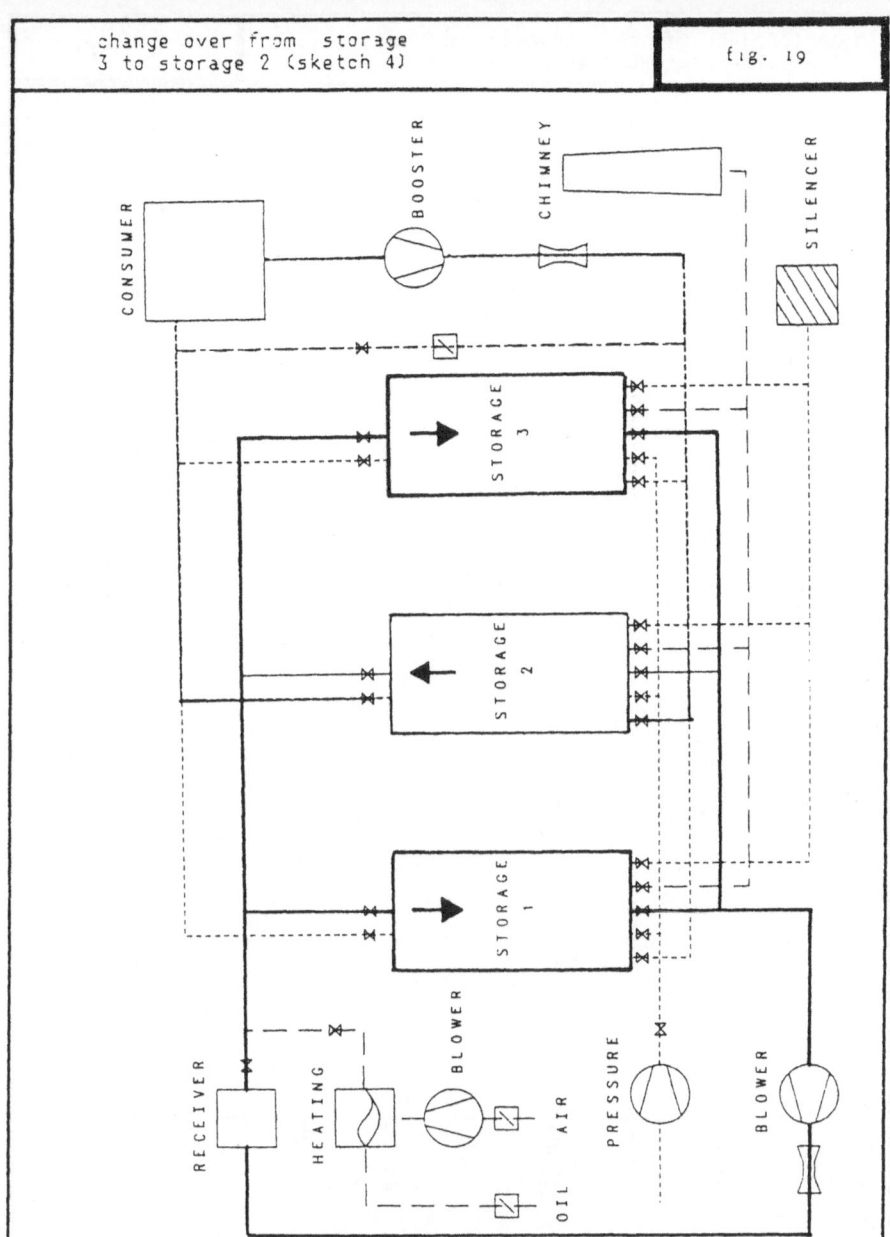

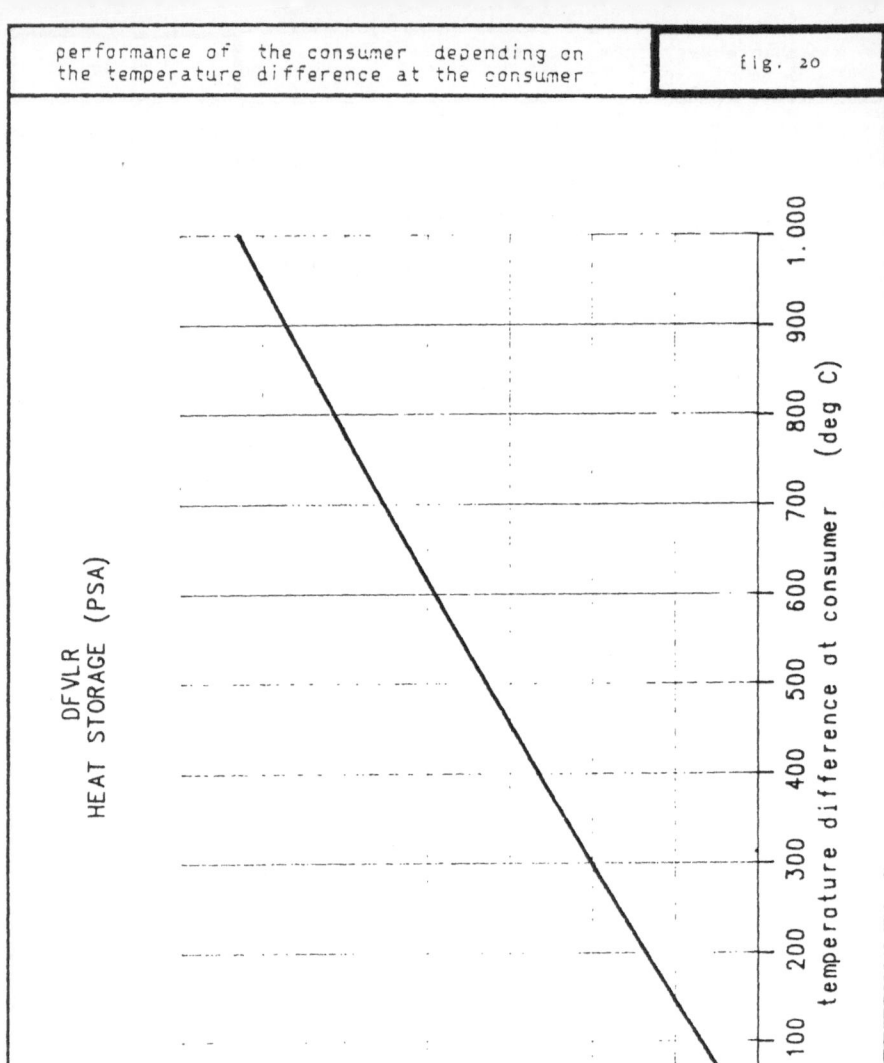

DFVLR
HEAT STORAGE (PSA)

temperature difference at consumer (deg C)

performance (kW)

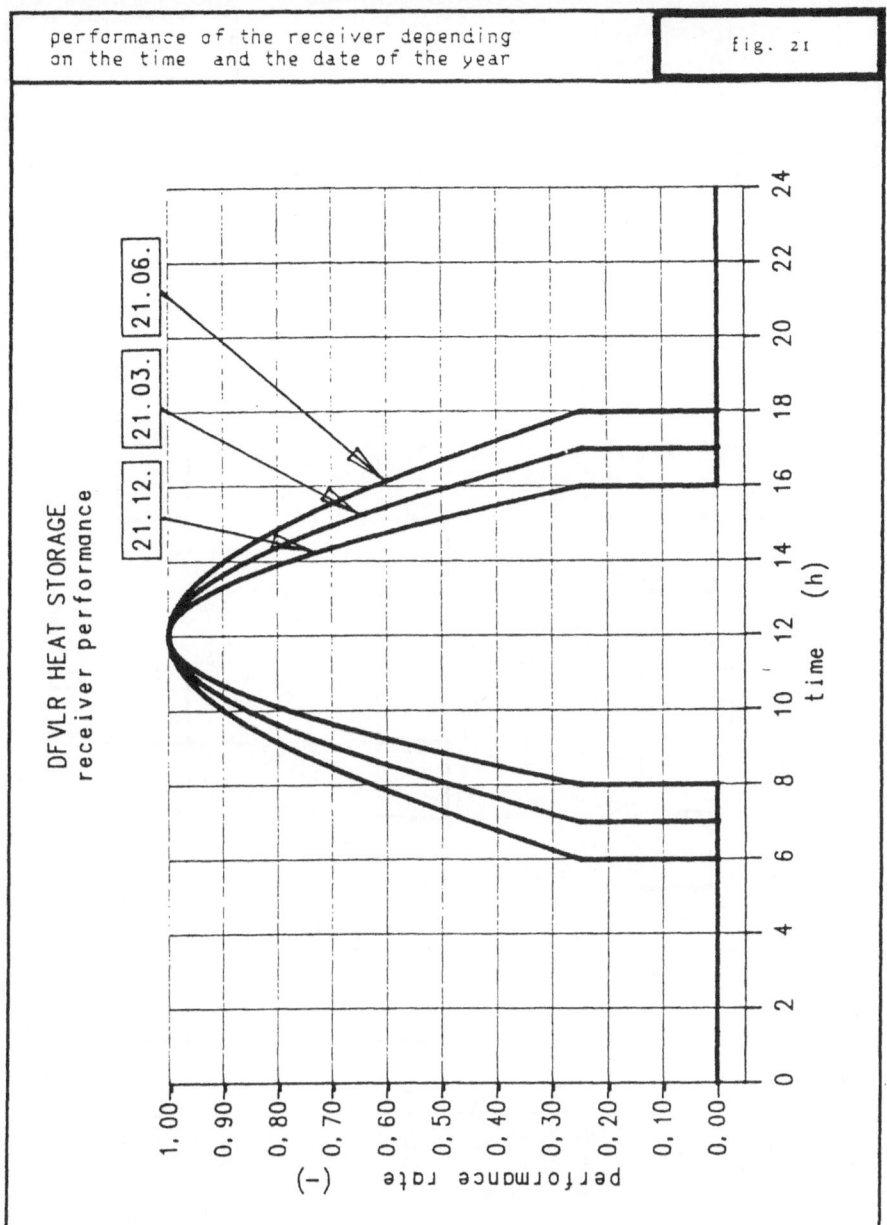

DFVLR HEAT STORAGE
receiver performance

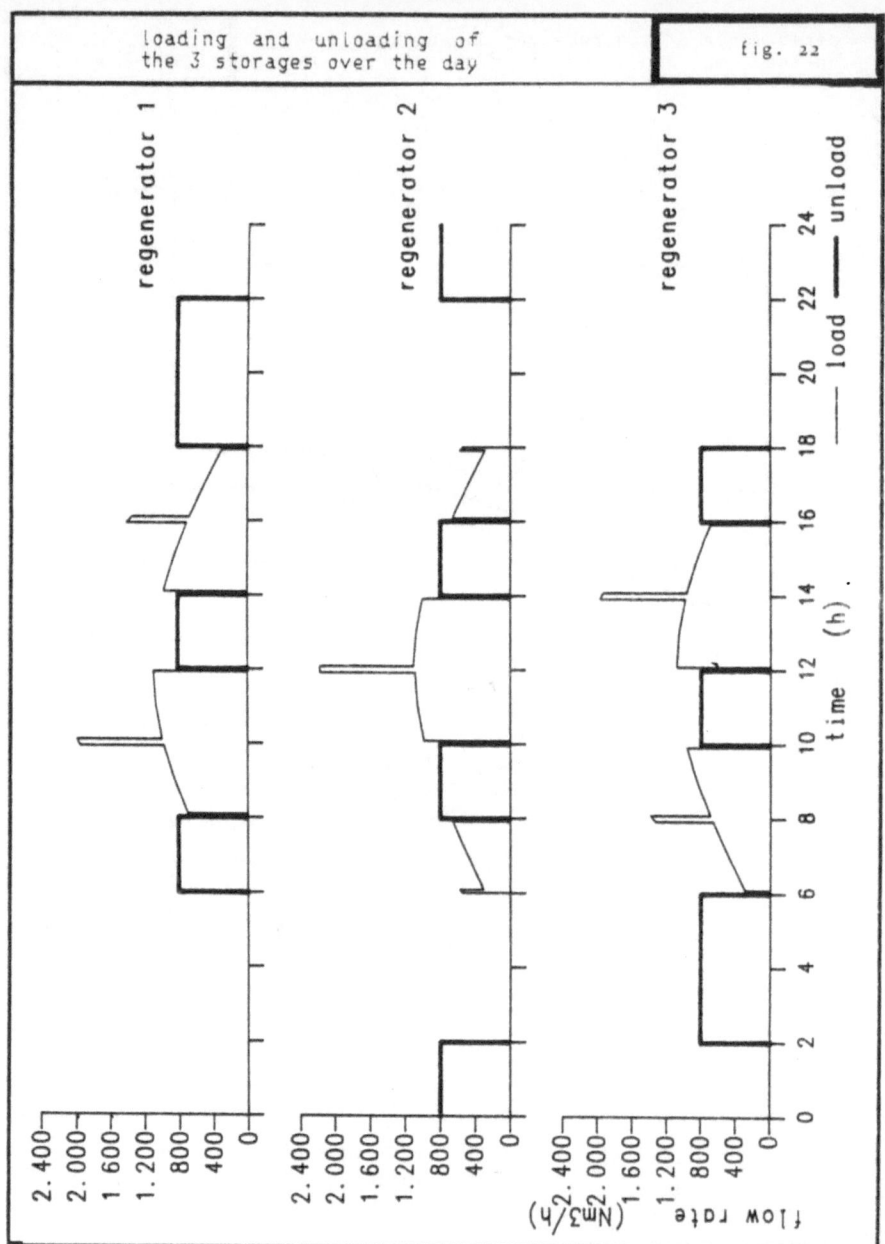

loading and unloading of the 3 storages over the day

fig. 22

- 174 -

alternative loading and unloading cycle | fig. 23

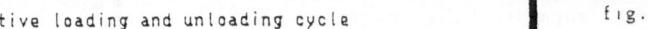

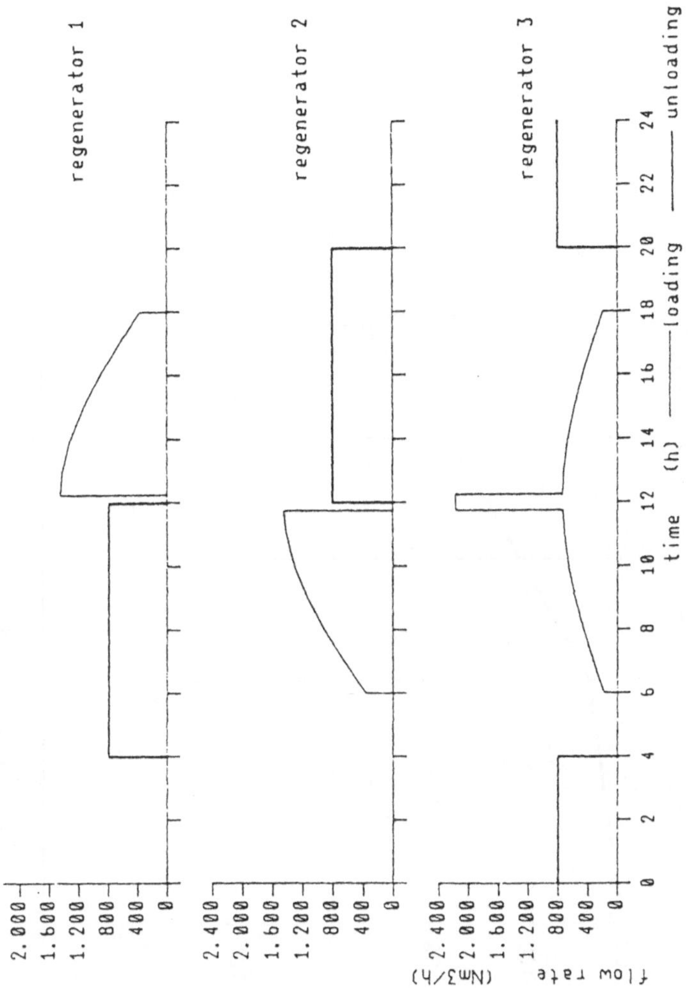

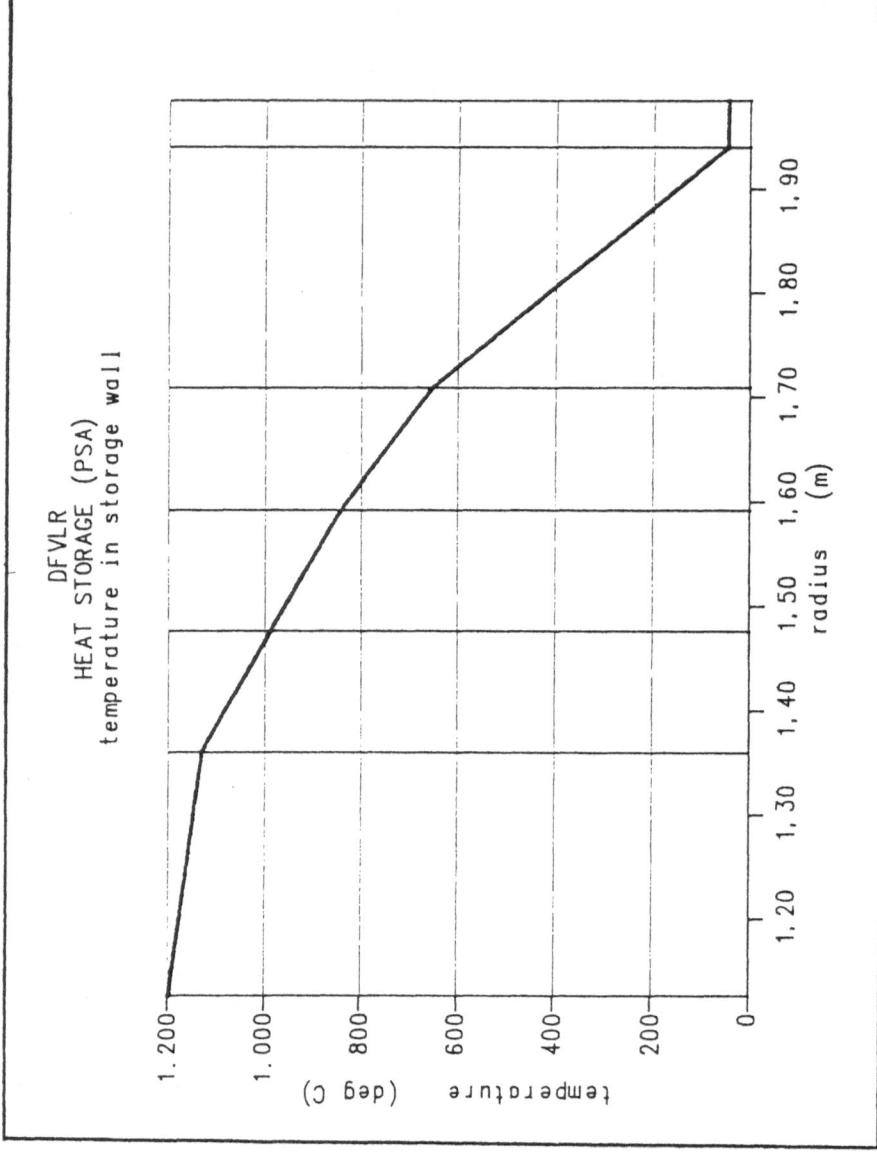

DFVLR
HEAT STORAGE (PSA)
temperature in storage wall

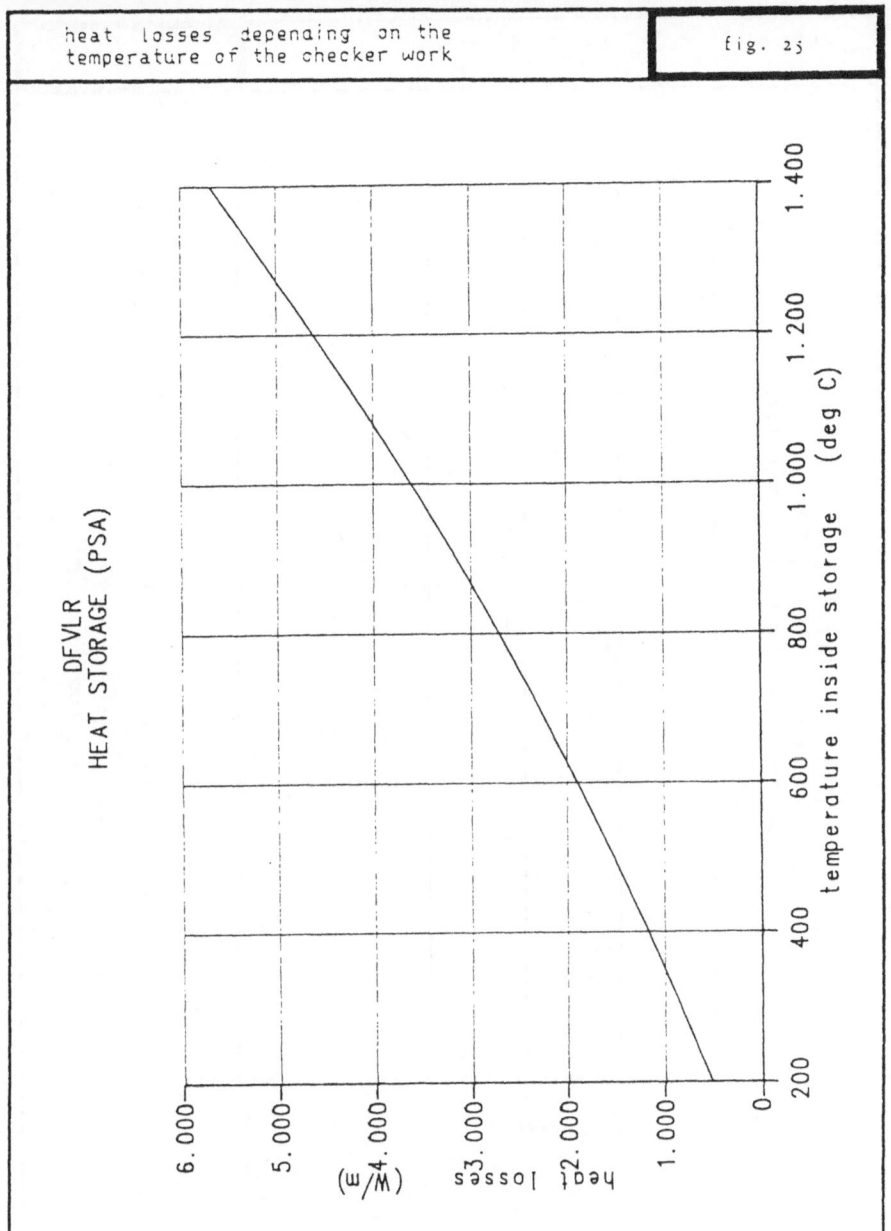

DFVLR
HEAT STORAGE (PSA)

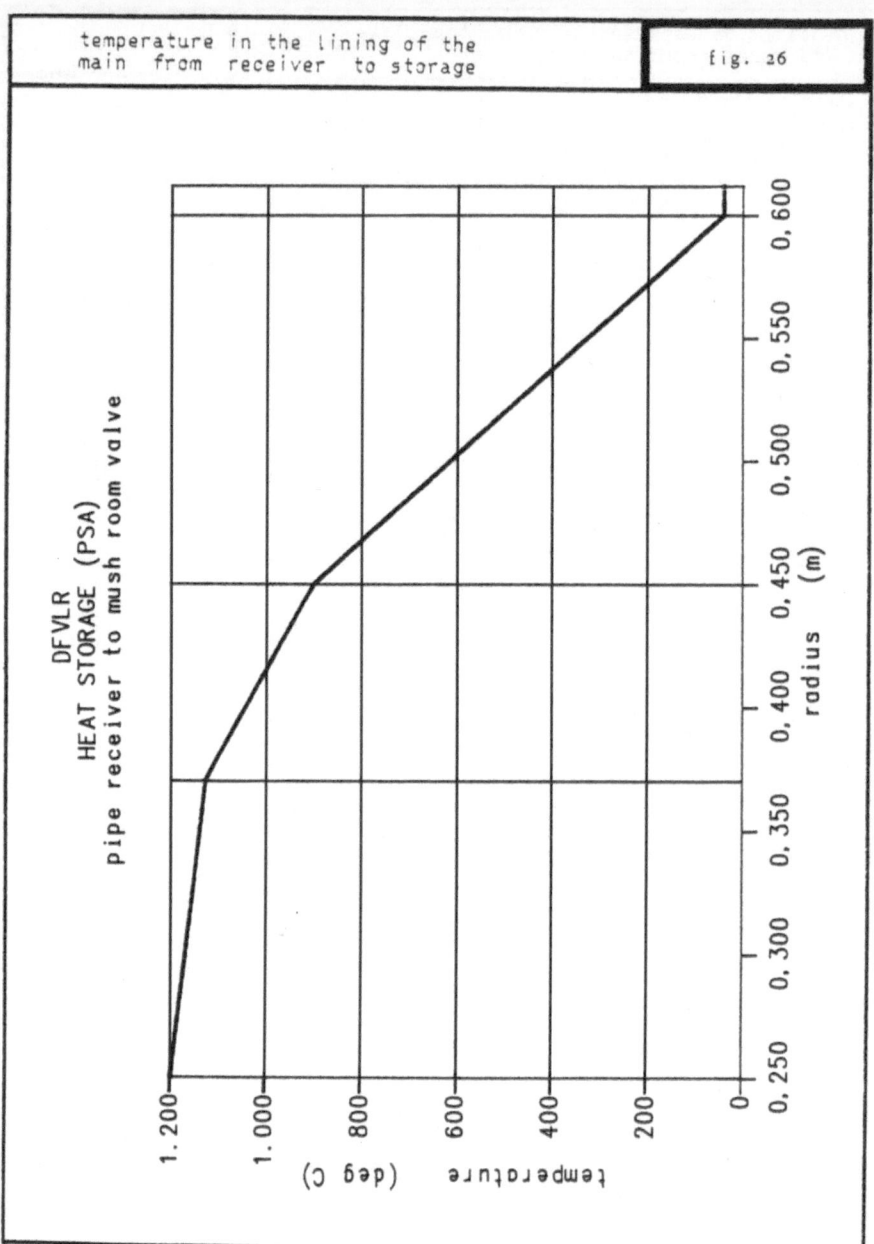

DFVLR
HEAT STORAGE (PSA)
pipe receiver to mush room valve

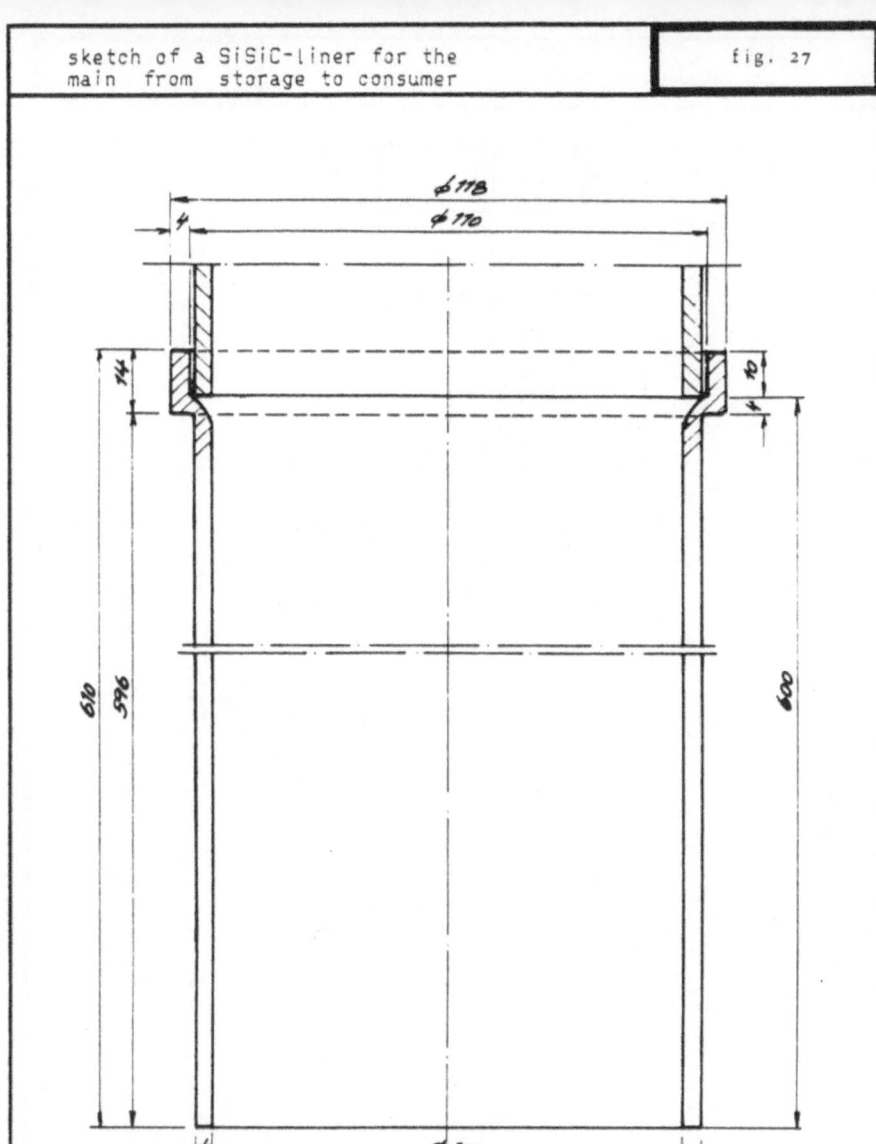

fig. 28

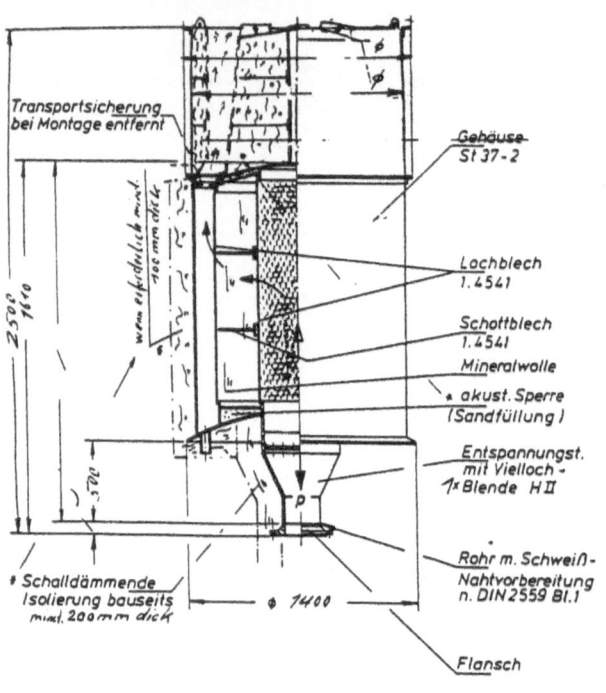

pressure in storage	20 bar
medium	air
temperature max	400 deg C
diameter of piping	50 mm
duration of pressure release	8 min
sound (50 m distance)	55 db(A)

heat exchanger for the consumer

fig. 29

DFVLR
HEAT STORAGE (PSA)
components: CONSUMER

no.	temp. deg C	app. volume
(1)	1000	800 Nm3/h air
(2)	200	300 Nm3/h air
(3)	800	1100 Nm3/h air
(4)	200	1100 Nm3/h air
(5)	170	5,32 kg/s oil
(6)	150	5,32 kg/s oil

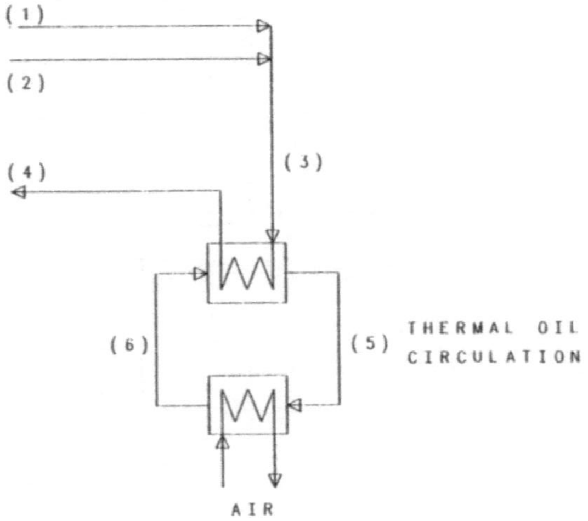
(1)
(2)
(3)
(4)
(5)
(6)
THERMAL OIL
CIRCULATION
AIR

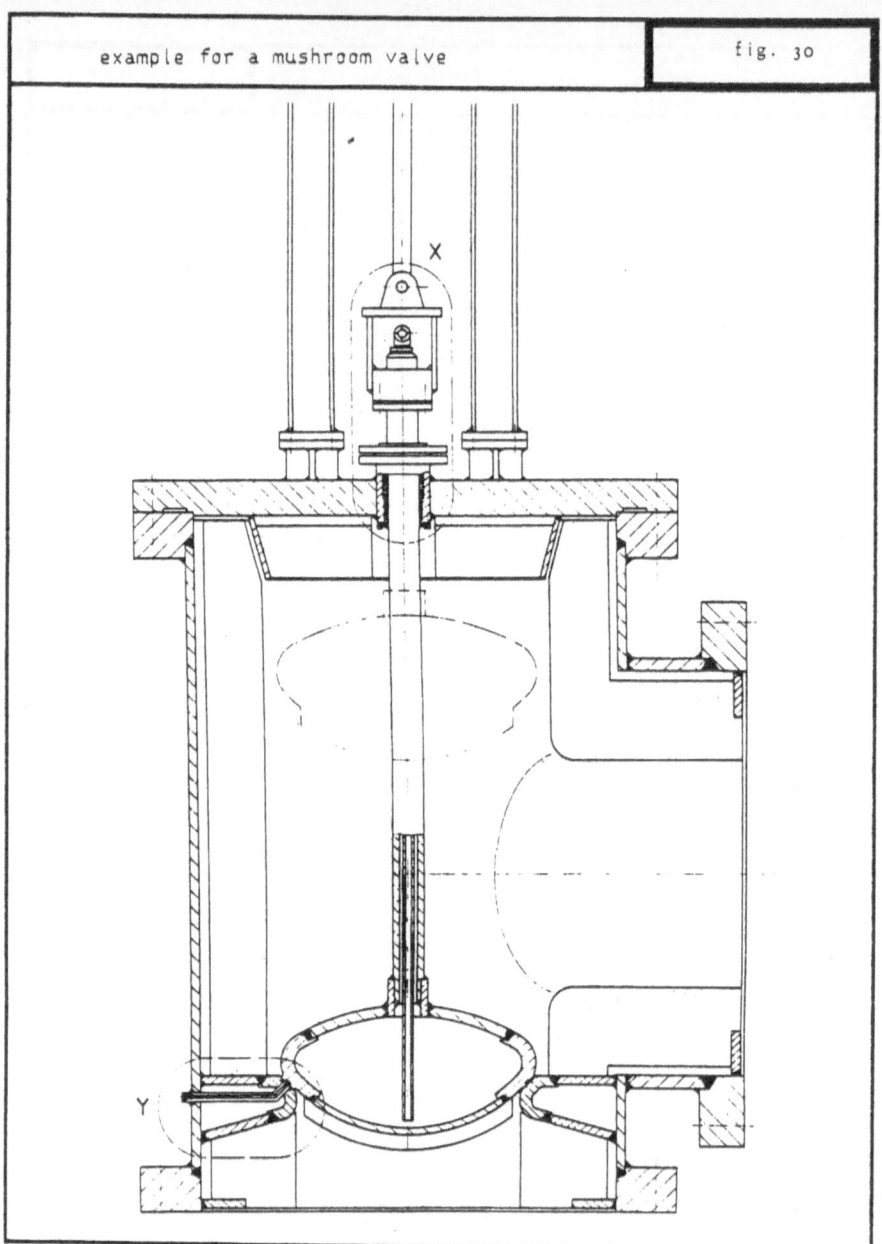

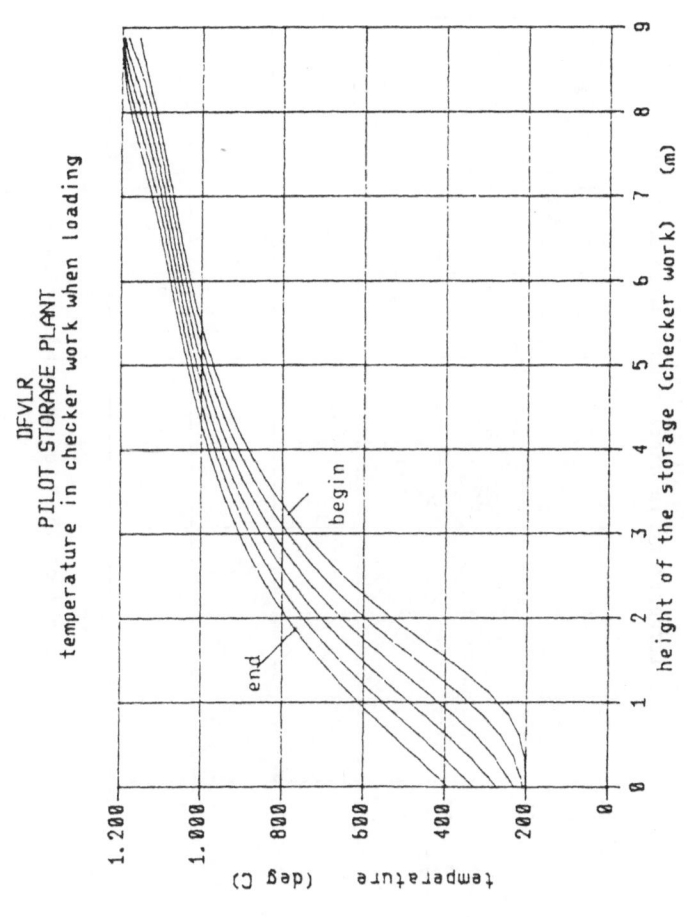

DFVLR
PILOT STORAGE PLANT
temperature in checker work when loading

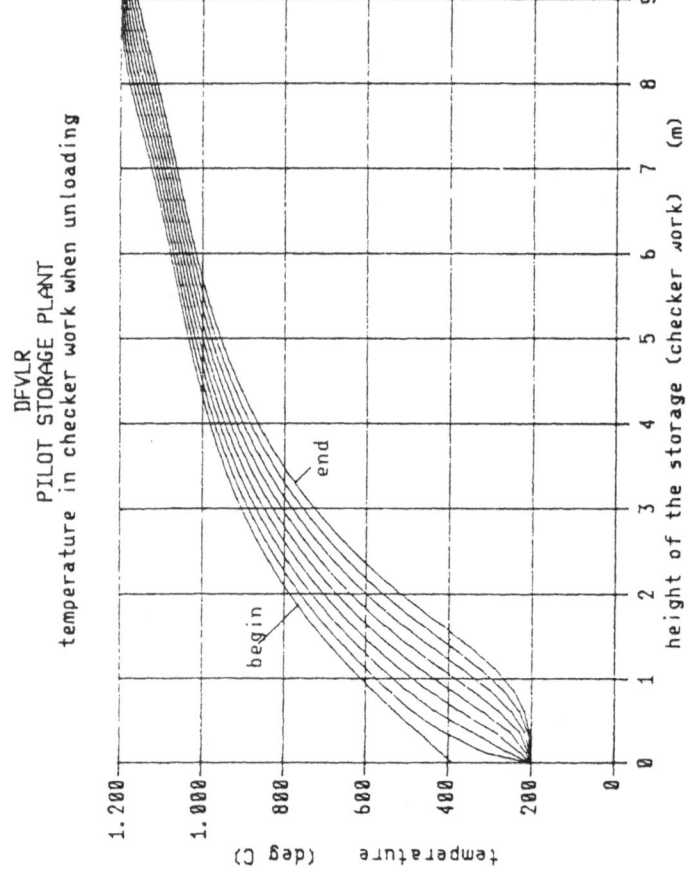

DFVLR
PILOT STORAGE PLANT
temperature in checker work when unloading

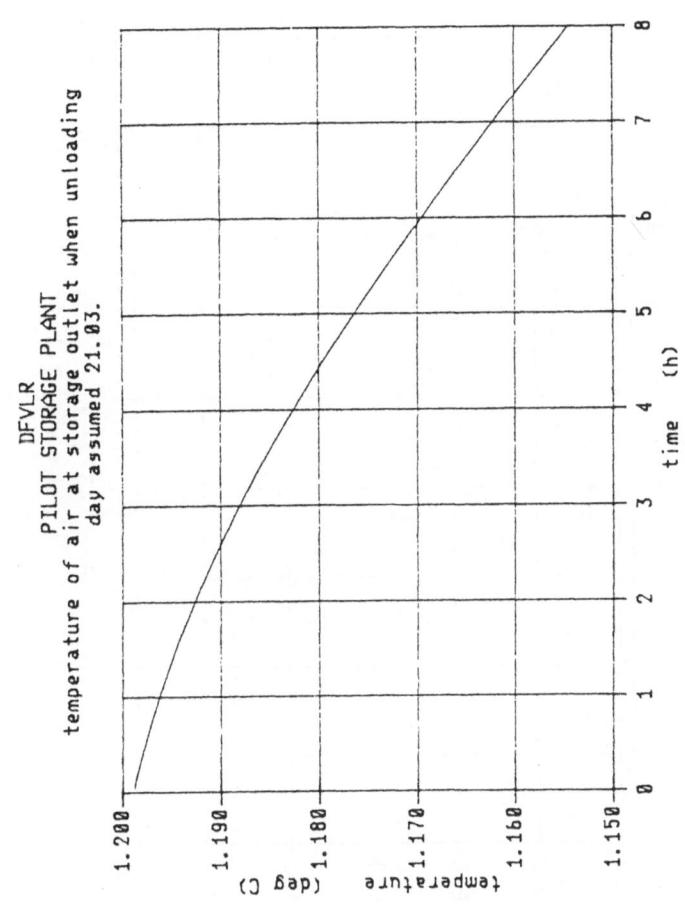

DFVLR
PILOT STORAGE PLANT
temperature of air at storage outlet when unloading
day assumed 21.03.

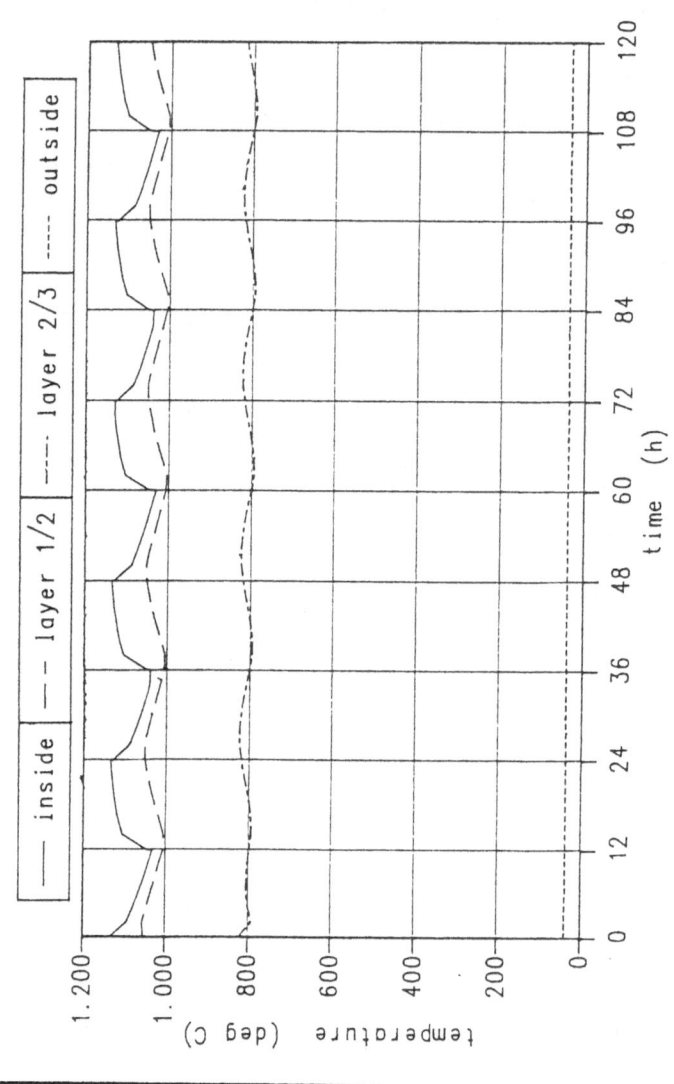

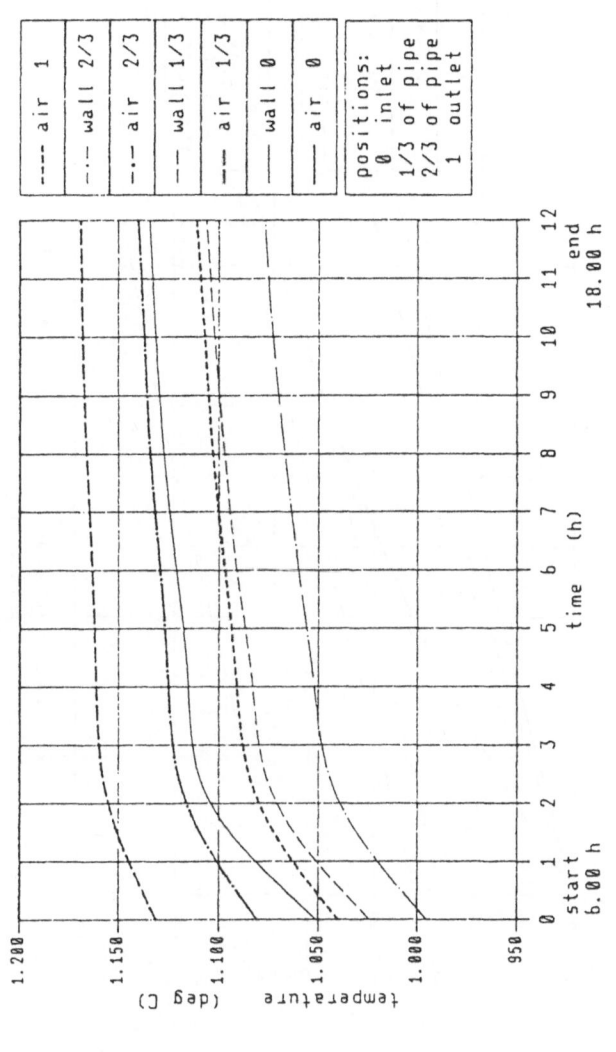

fig. 36

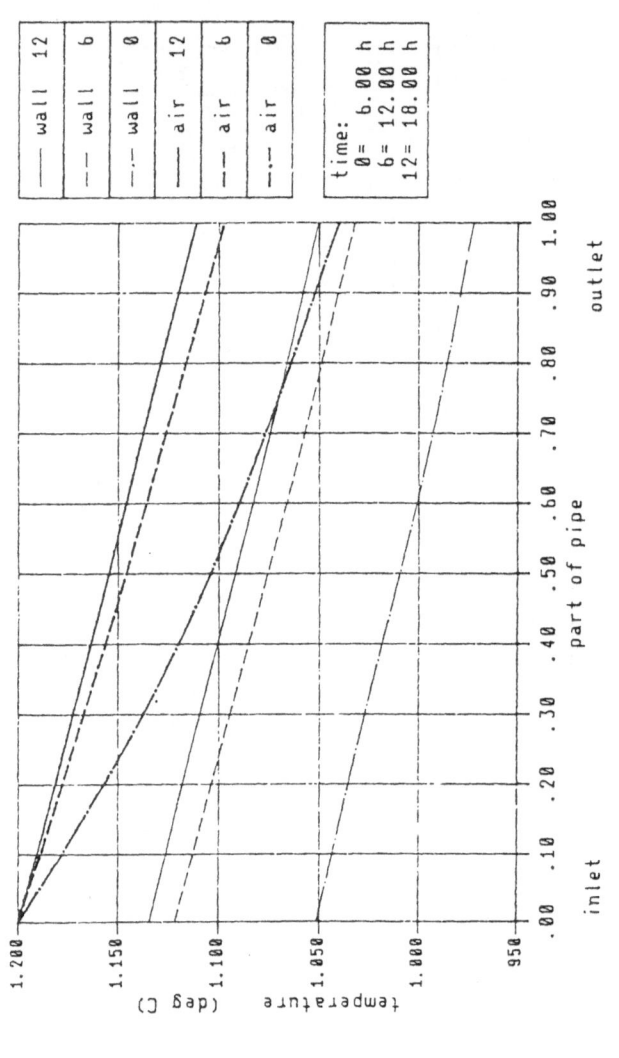

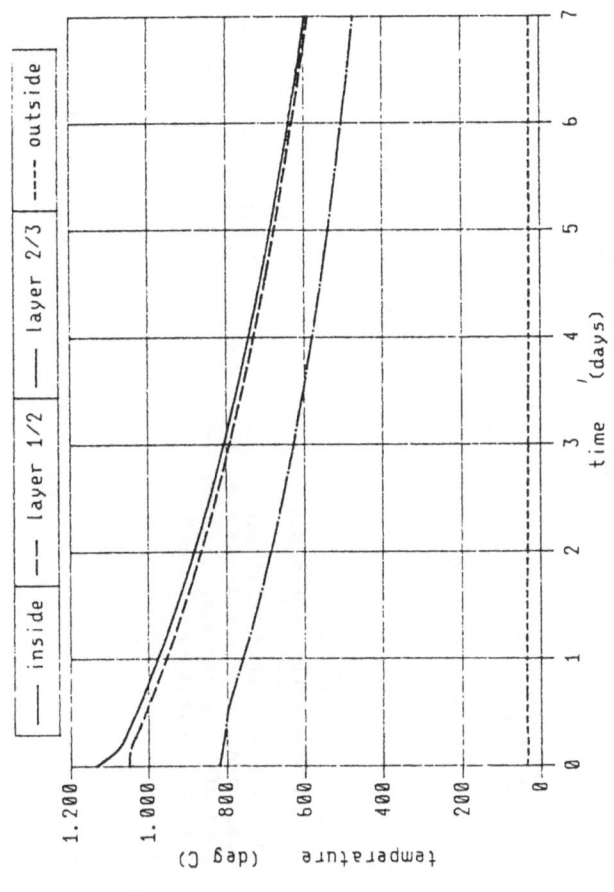

DFVLR
PILOT STORAGE PLANT
necessary opertion in case of disturbances

STORAGE-disturbance
===================

valves are not operating

----> load an other storage until
max. temperature of
supporting grid is reached

----> unload an other storage
until min. temperature for
the process is reached

effects for other components

The receiver can operate normally until a max. temperature at the
supporting grid is reached (400 deg C). Afterwards the receiver must
shut off. The consumer can work normally until a min. temperature
for the process is reached.

```
                    DFVLR
              PILOT STORAGE PLANT
     necessary operation in case of disturbances


RECEIVER-disturbance
====================

no energy from receiver to storage      ---->   shut off valve, wait
- short time                             ---->   shut off valve, wait, if
- longer time                                    necessary operating with
                                                 the heating system

the blower isn't working                 ---->   shut off receiver
                                                 shut off valve


effects for other components
----------------------------
The consumer is still working normally.

The latest loaded storage will be standby ---->  if necessary shut off
                                                 valve
```

DFVLR
PILOT STORAGE PLANT
necessary operation in case of disturbances

CONSUMER-disturbance
==================

the consumer can't accept energy or heat.
----> shut off booster
pressure release at storage
storage standby

the booster isn't operating, the pressure in storage can't increase
----> shut off consumer
as shown above

effects on other components

The receiver is working normally until a max. supporting grid
temperature is reachedt (400 deg C). Afterwards the receiver
must be shut off. storage will standby.

DFVLR
PILOT STORAGE PLANT
necessary operation in case of disturbances

GENERAL-disturbances
====================

electric   ---->   surching for the defect
controlling   ---->   surching for the defect,
                             operating by hand
power failure   ---->   if necessary operating with
                             a emergency power unit,
                             cooling of valves is
                             necessary

effeccts for other components

If necessary other components (shut off from power) must be shut off.
If wished the consumer can be operate with the emergency power unit.

- 193 -

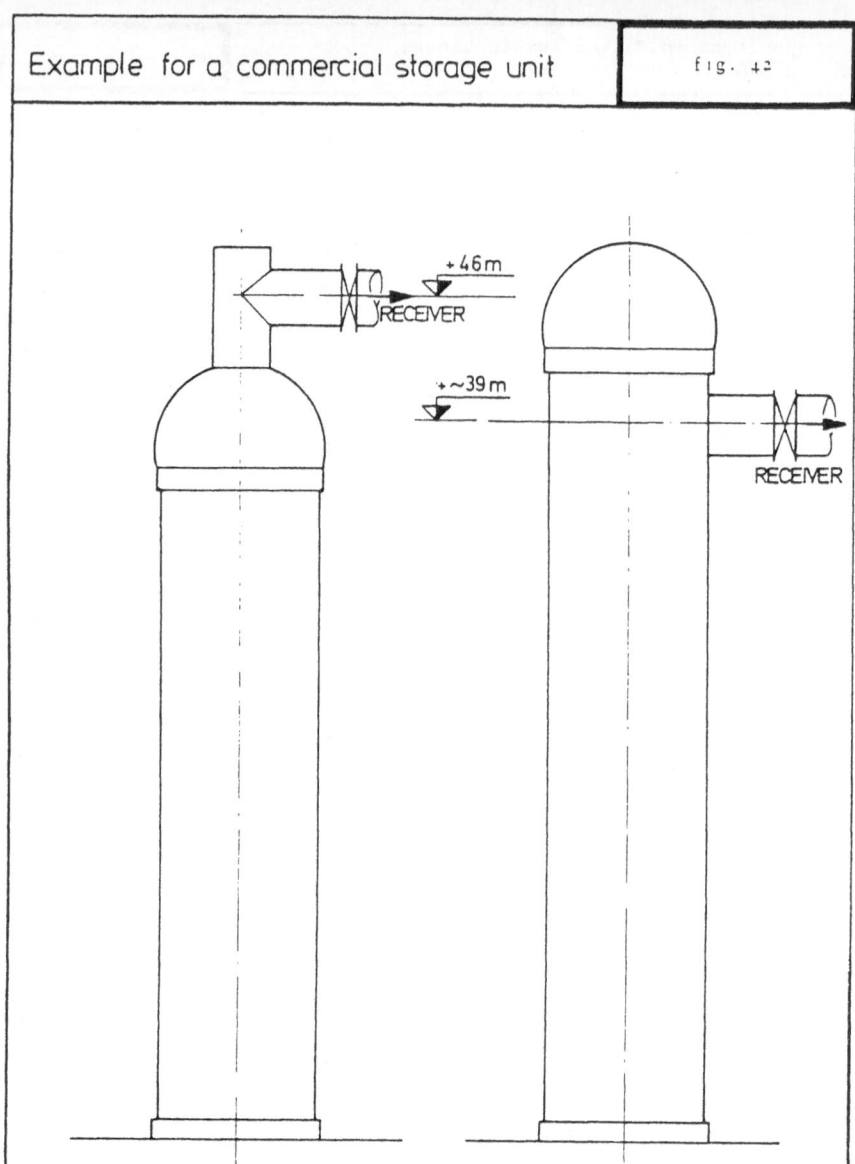

+ 46 m
RECEIVER

+ ~39 m

RECEIVER

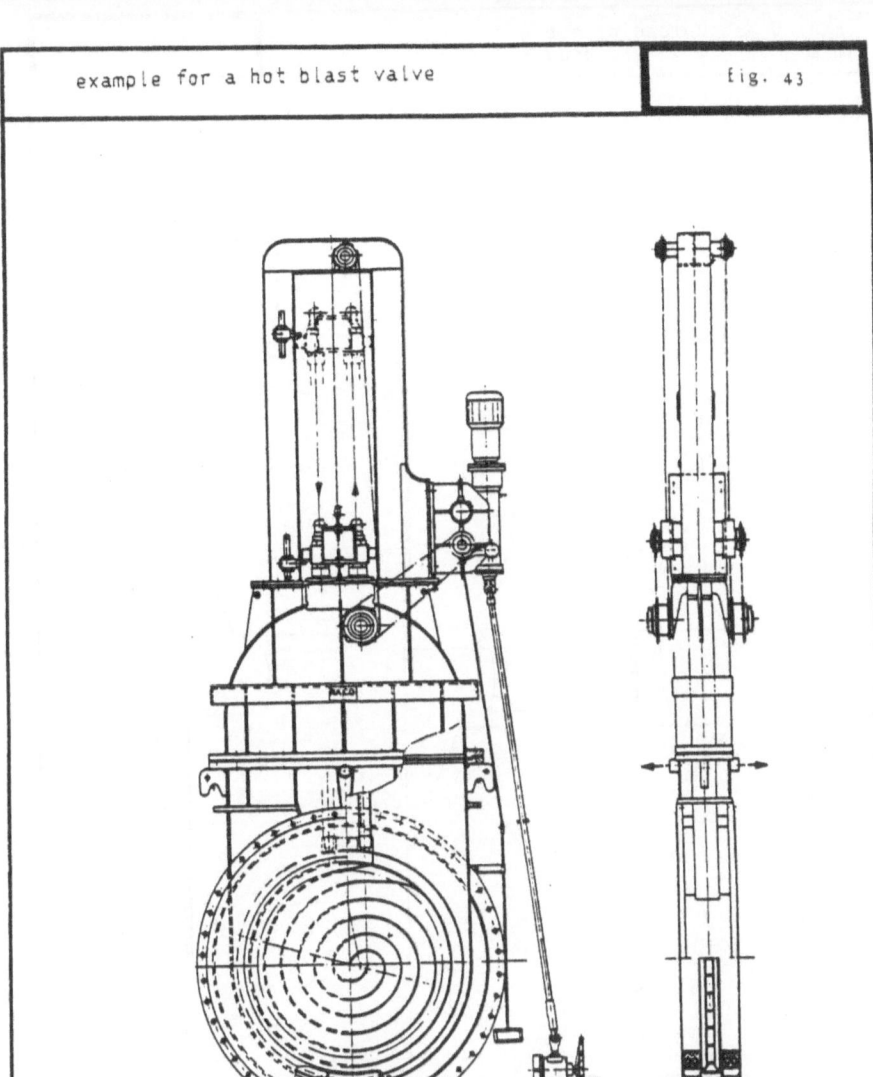

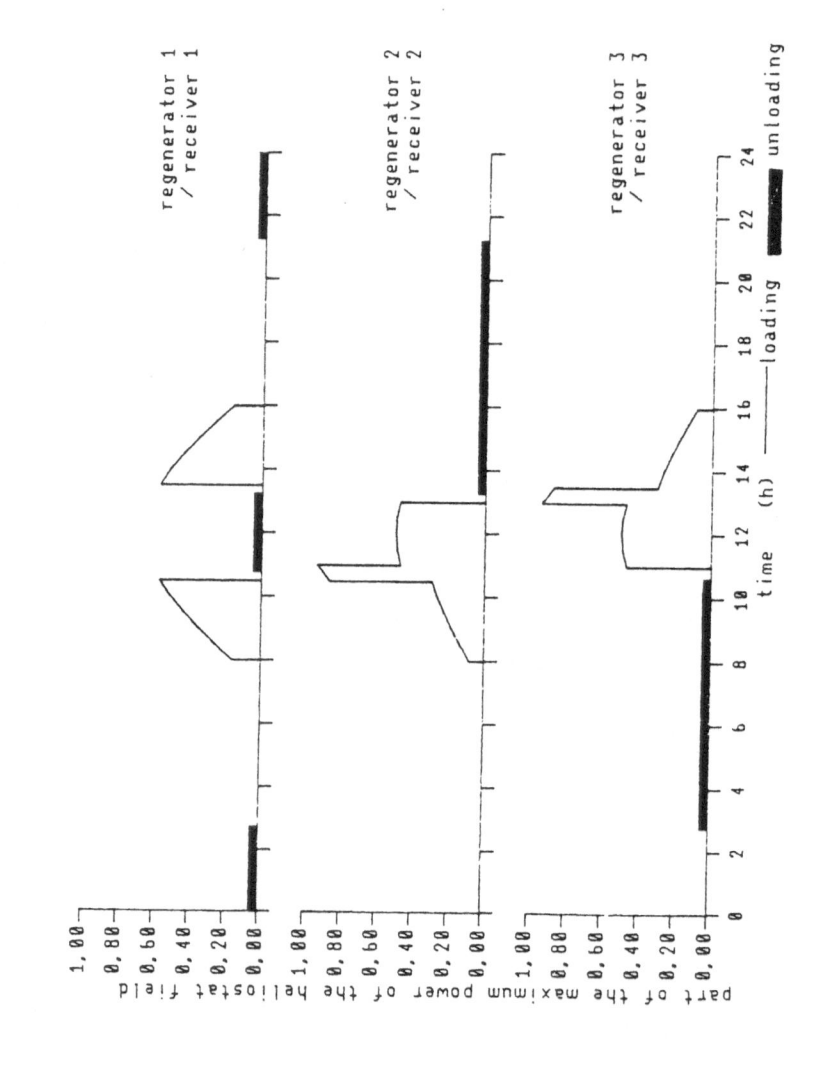

loading and unloading cycle of a C S P with 3 storages

fig. 44

- 196 -

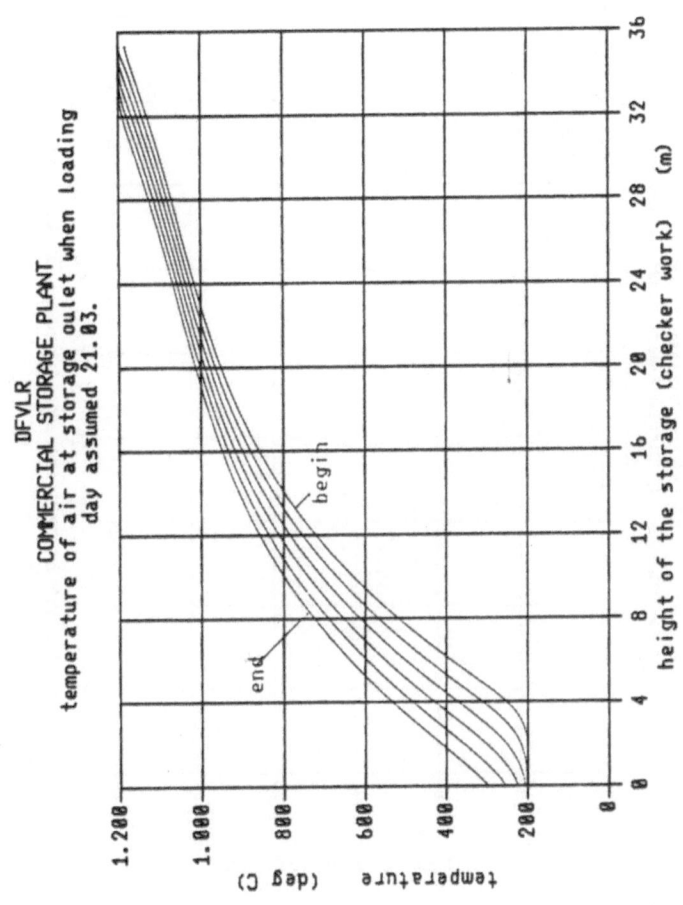

DFVLR
COMMERCIAL STORAGE PLANT
temperature of air at storage oulet when loading
day assumed 21.03.

- 197 -

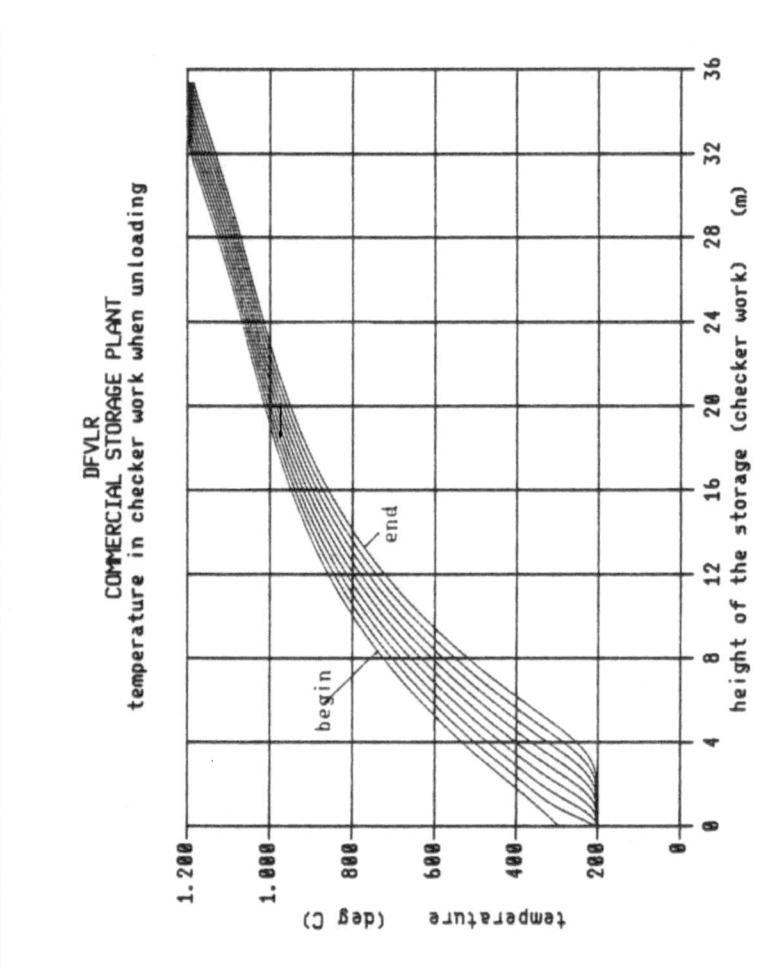

DFVLR
COMMERCIAL STORAGE PLANT
temperature in checker work when unloading

end

begin

height of the storage (checker work) (m)

temperature (deg C)

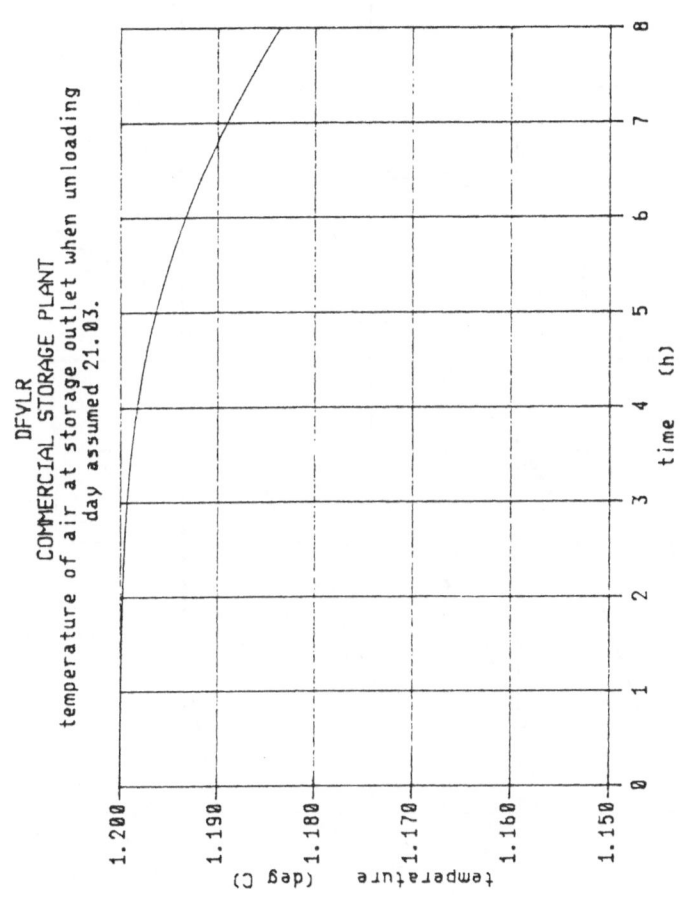

DFVLR
COMMERCIAL STORAGE PLANT
temperature of air at storage outlet when unloading
day assumed 21.03.

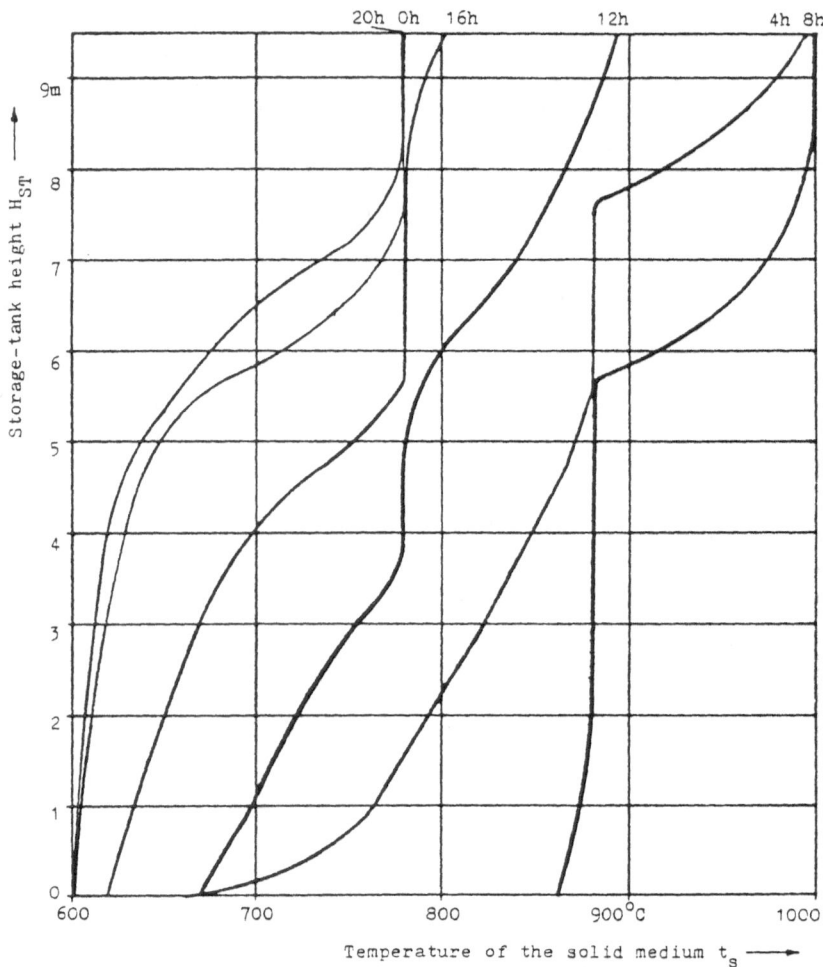

Fig. 13: Temperature profiles in the storage-tank

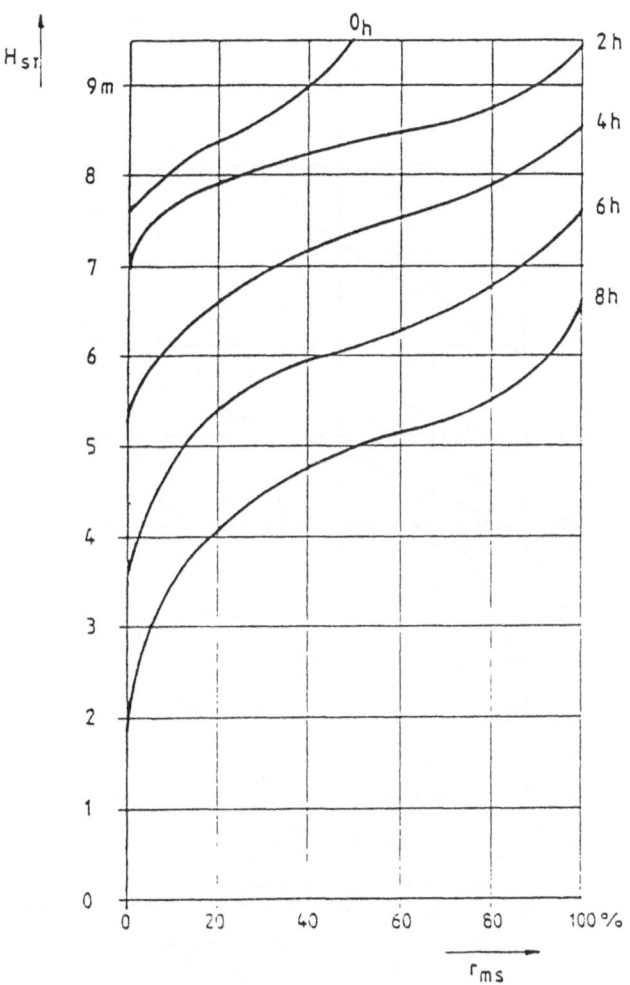

Fig. 14 a Melted salt share r_{ms} versus storage height H_{ST}
(charging cycle)

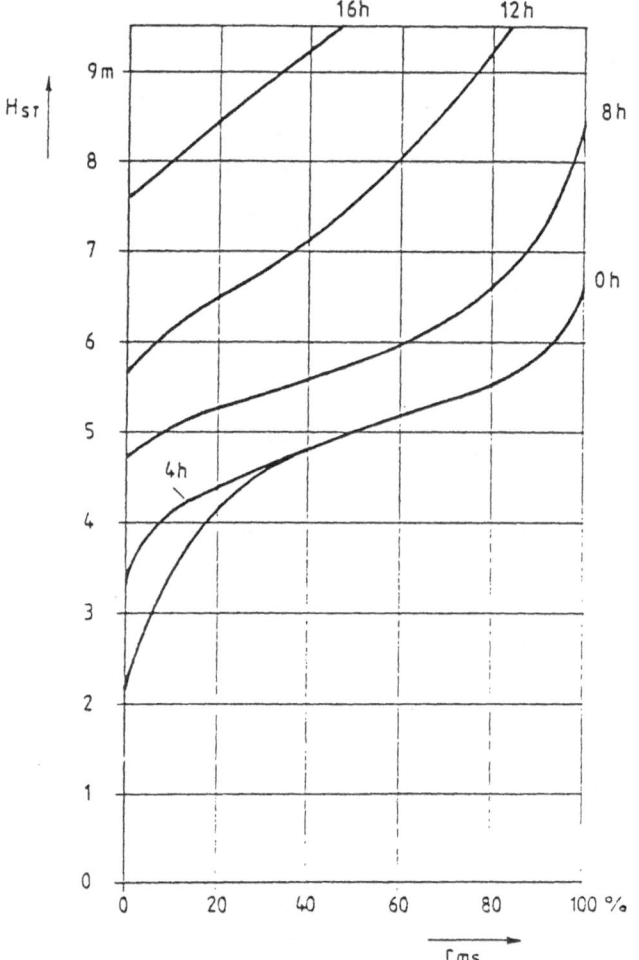

Fig.14b Melted salt share r_{ms} versus storage height H_{ST}
(discharging cycle)

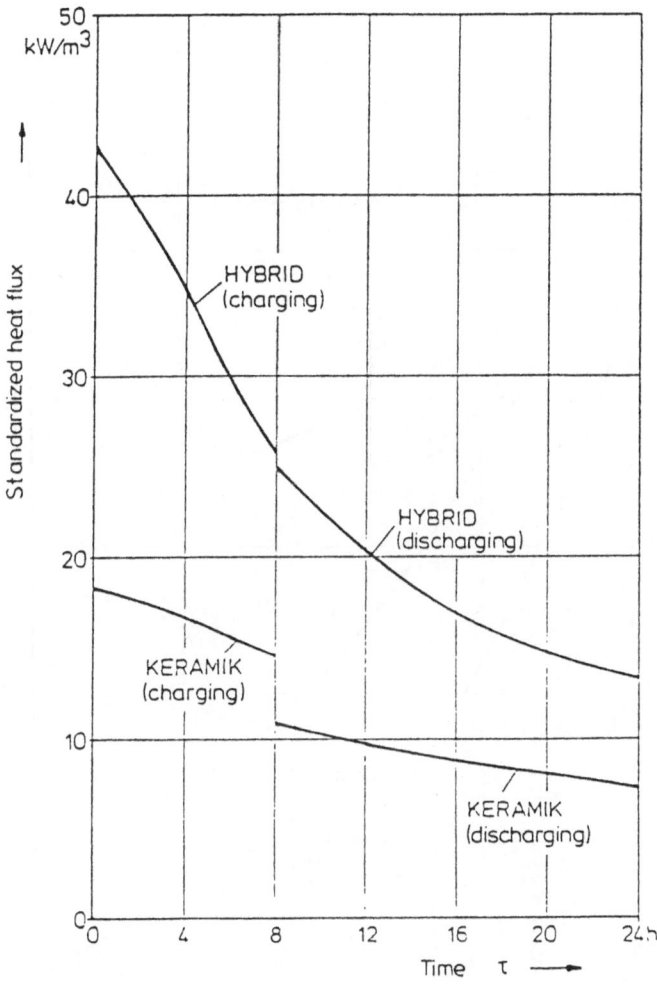

Fig.15 Average transferred heat flux related to storage volume for ceramic - and hybrid - relining

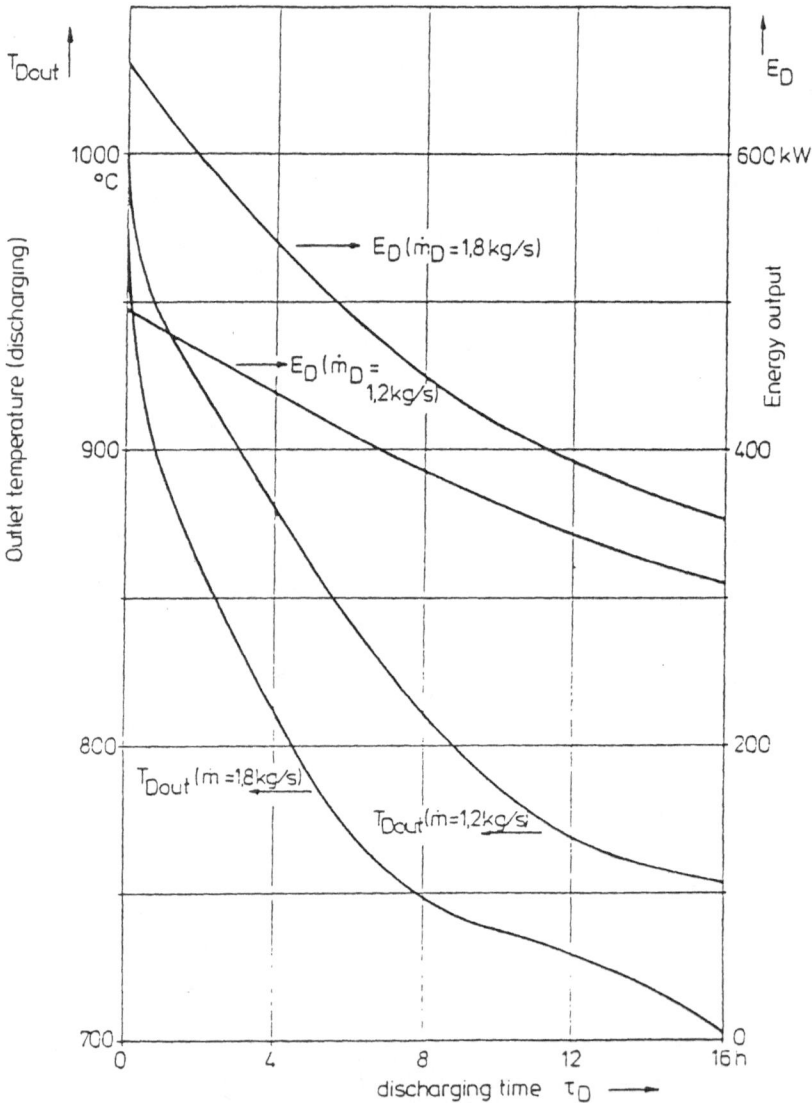

Fig.16   Timecurves of outlet temperature and of energy output

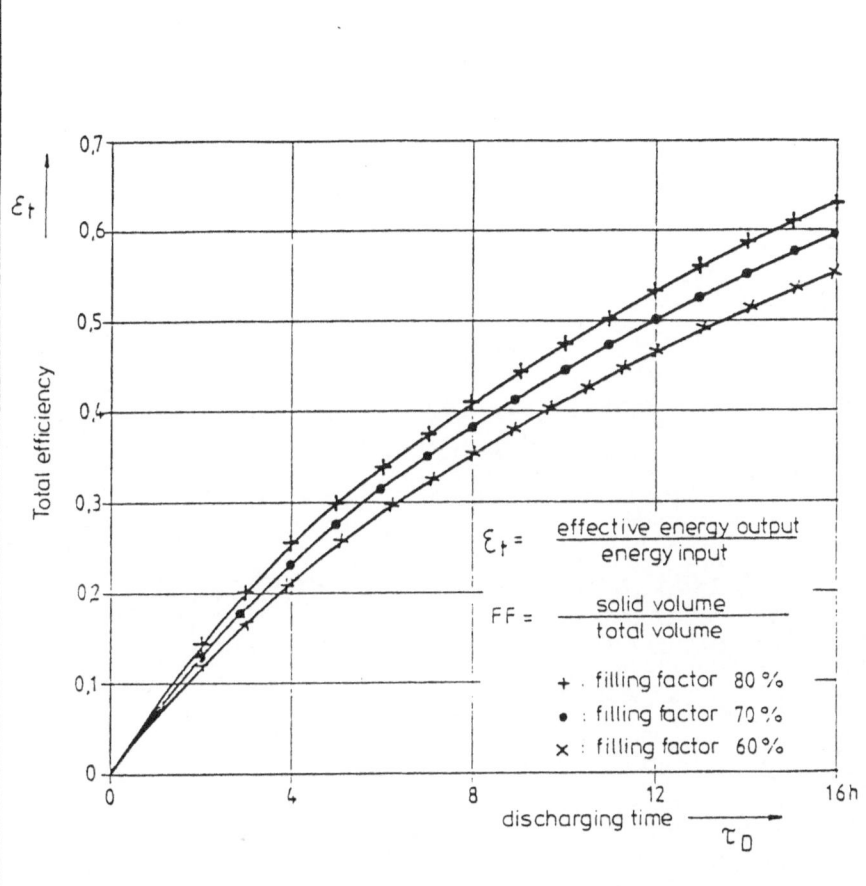

$$\mathcal{E}_t = \frac{\text{effective energy output}}{\text{energy input}}$$

$$FF = \frac{\text{solid volume}}{\text{total volume}}$$

+ : filling factor 80 %
• : filling factor 70 %
× : filling factor 60 %

Fig. 17 Timecurve of the total efficiency $\mathcal{E}_t$ for different filling factors FF

8   DRAWINGS

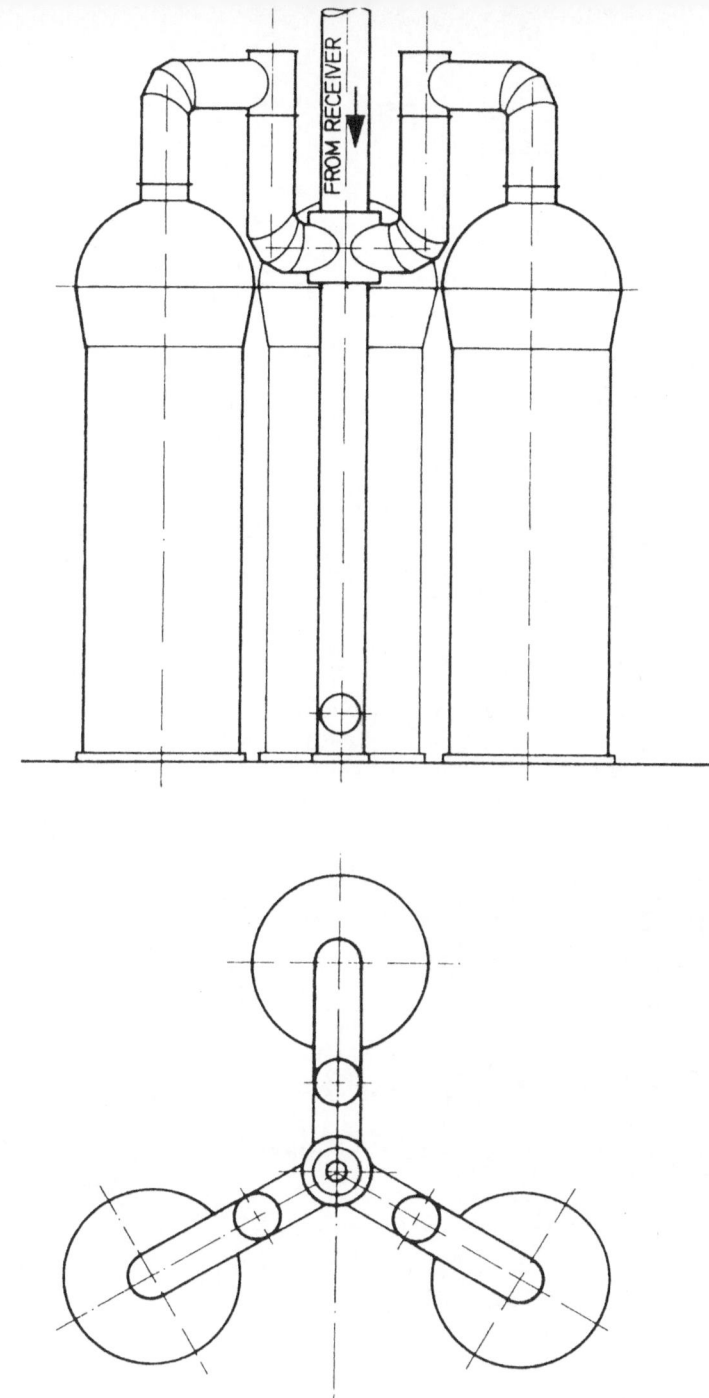

FROM RECEIVER

HIGH TEMPERATURE STORAGE PLANT
TRIANGLE POSITION

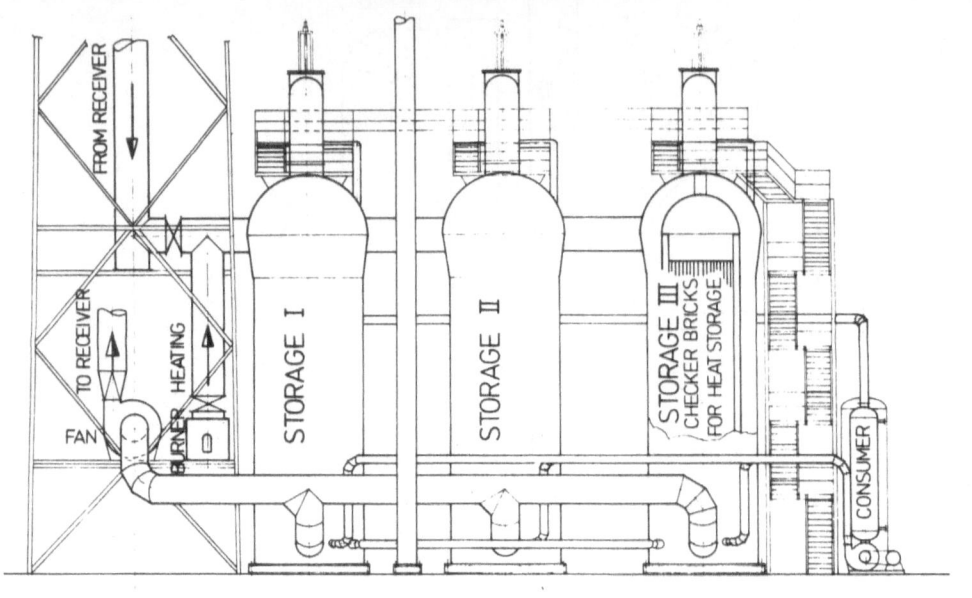

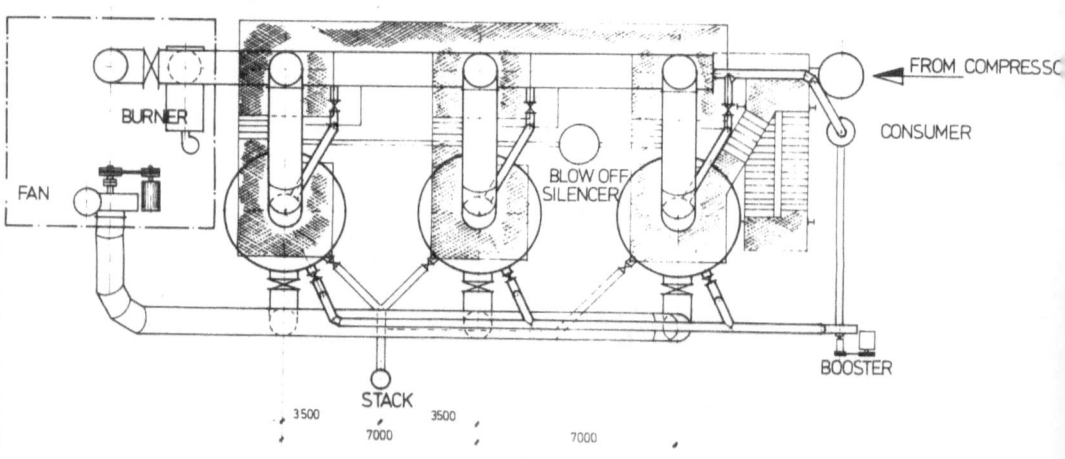

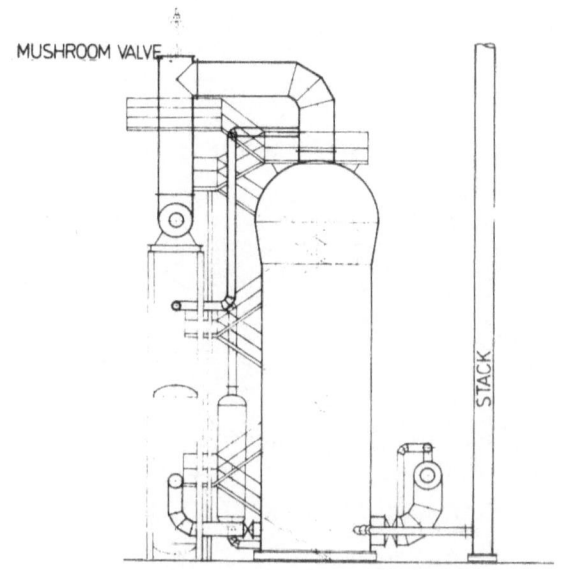

MUSHROOM VALVE

STACK

HIGH TEMPERATURE STORAGE PLANT

EXPERT OPINION AND CO-OPERATION

IN THE DEVELOPMENT PROGRAM

HIGH-TEMPERATURE-STORAGE-TANK

TH.J. BOHN

K. WERNER

W. BITTERLICH

F.J. JOSFELD

UNIVERSITÄT ESSEN - GHS

ENERGIE - UND KRAFTWERKSTECHNIK

EKT

Contents

1.	Nomenclature	215
2.	Introduction	217
3.	Objective	220
4.	Principle Structure of the System	221
5.	Principle Structure of the HTST	222
6.	Basic Thermodynamical Considerations	224
7.	Consideration of the Phase-Change	232
8.	Characteristic Material Data	234
9.	Characteristic Design Data of the System	237
10.	Way of Optimizing	241
11.	Results	246
12.	Conclusion	250
13.	Figures	251
14.	References	259
15.	Appendix	260
	Numerical Output of the Simulation Program	261
	Numerical Output No. I	263
	Numerical Output No. II	274
	Numerical Output No. III	285
	Numerical Output No. IV	296
	Numerical Output No. V	307

NOMENCLATURE

A	surface, cross-sectional area	m^2
c	specific heat	$J/(kg \cdot K)$
d	diameter	m
D	diameter	m
E	energy	J
FF	filling-factor	$\%$
g	acceleration by gravity	m/s^2
H	height	m
$\dot{H}$	enthalpy flux	J/s
Δh_f	heat of fusion	J/kg
k	combined heat transfer	$W/(m^2 \cdot K)$
L	length	m
m	mass	kg
$\dot{m}$	mass flow	kg/s
N_c	number of flow canals	$-$
Nu	Nusselt-Number	$-$
p	pressure	Pa
Pr	Prandtl-Number	$-$
$\dot{Q}$	heat flux	W
Re	Reynolds-Number	$-$
r_{ms}	share of melted salt	$\%$
s	web-width between the flow canals	m
t	temperature	$^{\circ}C$
T	temperature	$K, ^{\circ}C$
u	specific inner energy	J/kg
U	circumference	m
v	velocity	m/s
V	volume	m^3
x	flow path	m
y	transformed co-ordinate	m
z	transformed co-ordinate	s
z_f	geodesic height of the fluid	m

α	heat transfer coefficient	$W/(m^2 \cdot K)$
ε	efficiency	-
φ	angle of co-ordinate transformation	$^{\circ}$
ϕ	auxiliary function	-
λ	coefficient of thermal conductivity	$W/(m \cdot K)$
ν	kinematic viscosity	m^2/s
ρ	density	kg/m^3
τ	time	s

SUBSCRIPTS

amb	ambient
c	crystallization
C	charge
cf	heat conduction in x-direction / fluid medium
cs	heat conduction in x-direction / solid medium
crit	critical
D	discharge
f	fluid
f/s	heat transfer fluid - / solid medium
i	inner
in	inlet
m	melting
max	maximum
min	minimum
out	outlet
Qf	cross-section / fluid
red	reduced
s	solid
ST	storage (tank)
t	total

Introduction

In order to achieve a permanent availability and an assimilation of energy offer and demand for solar power plants, in principle one can go two different ways.

First one can integrate a second, for example a fossil, source of energy into the existing cycle, so that periods of less radiation can be spanned and start-up procedures can be improved.

Otherwise these advances in efficiency can be rendered possible by an energy storage system.

If the storage is designed in a way that the requirement of heat for the linked process can even be covered in periods with less sunshine, one gets the advantage, that the solar plant is able to work independent of other energy sources.

With regard to thermal storage there are mainly five parameters, which determine the capacity of the energy stored:

- the volume of the storage tank
- the density of the storage medium
- the specific heat of the storage medium
- the difference between the maximum temperature of the charging cycle and the minimum required temperature for the linked process
- the heat of fusion.

Mostly the enlargement of the storage volume is
opposed to technical and economic reasons.

The next two parameters, density and specific heat,
are depending on the choice of material and
consequently they are limited.

The fourth value, the maximum storage temperature is
theoretically unlimited, the economy of operation
however demands, that the level inside the storage is
as near as possible to the level of the useful
process.

Moreover most applications need a rather constant
offer respecting the working fluid, so that the
outlet-temperatures of the discharging-cycle should
also change very little.

These demands can very well be fulfilled by thermal
storage tanks, which use the change of phase of the
storage medium and thereby the heat of fusion.

The reasons, which are chargeable to the rare
application of storing latent thermal energy are
mainly the following:

- the segregation of the material
- the inhibition of the heat transfer by local phase
  changes
- no ideal melting or crystallizing; undercooling or
  superheating necessary
- aggressive behavior against surrounding materials.

By using a concept, in which the pure phase-change-medium is equally distributed in the solid matrix, it shall be ascertained, whether the above mentioned disadvantages of the conventional storage of latent thermal energy can be suppressed.

A combination of ceramic as the basic material and of incorporated salts as the phase-change-medium is investigated.

So the material combines the storage of sensible and latent thermal energy.

This new concept of a High-Temperature-Storage-Tank (HTST) will here theoretically be investigated and compared with conventional solutions.

Objective

The aim of the study was to develop a calculation
routine being able to simulate the performance
characteristic of a HTST.
This simulation program then could be used to achieve
informations about the following three main criteria:

- the optimization of the geometry of the active
  storage part
- the optimization of the filling-factor
  (ratio solid-/gas medium)
- the comparison of the different relining materials
  ceramic and ceramic/salt-combination.

Especially the third investgation point renders
possible to draw conclusions from the use of the so
called Hybrid-material, which is able to store as well
sensible as latent heat.
In general the whole examination aimes at the
determination of the in- and outlet energy flow rates
and at the occuring profiles of temperature in the
storage.
Finally one can deduce the efficiencies from these
calculated values and can compare the given concept
with other storage systems.

Principal structure of the system

In order to enable a 24 hour-performance for the
linked consumer or energy conversion system without
using a second energy source besides the radiation of
the sun the high-temperature-storage-tank should be
integrated in a system as shown in fig.1 and fig. 2.
The receiver should render possible to charge the
high-temperature-storage-tank and at the same time one
of the regenerators of the planned three stove
operation system (fig.1).
As the only task of the regenerators is to accomodate
the pressure of the gas flow to the demanded value of
the consumer, one can use conventional components,
which are already successfully operated in the heavy
industry.
After being charged the high-temperature-storage-tank
can be discharged by the outer circle as shown in
fig.2.
The capacity of the receiver has to be big enough to
charge continously one hot blast stove, in order to
enable the daytime operation and to charge the HTST
for the nighttime operation of the plant.
As only the HTST is the subject of this study, we
won't discuss more details of the whole system in this
report.

Principle structure of the HTST

The development of a HTST can be supported by
extensive and of long standing experiences of the
steel- and iron-trade.
This know-how is especially related to the following
four fields:

- the building of the pressure vessels
- the thermal isolation of the active storage part
- the high-temperature resistant stonework of the dome
- the forms of the relining bricks related to thermal
  expansion and problems of mechanical stability.

As fig.3 shows there are existing hot blast stoves,
which use an outside combustion chamber.
Especially by looking at the arrangement under B,
where stove and combustion chamber have separated
domes, one can easily imagine, that the combustion
chamber can be substituted by the solar receiver
system.
Fig.4 represents the temperature profile in a
stonework of a high-power hot blast stove.
A temperature difference of about 1500 K can be
reduced over a length of 700 mm even with maximum
temperatures of 1650 $^{\circ}$C, if different layers of
corresponding materials are used.
In the highest temperature range often Silika-stones
are employed, which possess an excellent behavior

related to thermal expansion between 600 °C and 1450 °C, as shown in fig.5.

One of the typical relining forms today used in hot blast stoves is pictured in fig.6 and fig.7.

The hexagonal flow canals are arranged vertical in every hexagonal stone.

The checkers are placed in a so called "tongue and groove" manner, where each of the three flues at the bottom of a brigg straddles three bricks in the course below.

Combined in such a way the checkers leave a gap around their perimeter, so that problems of mechanical stability and thermal expansion can be reduced to a minimum /3/.

Whereas the preceding technologies can be regarded as the state of art, there are mainly three remaining problems, which still have to be solved:

- the technological and the chemical treatment of the Hybrid-Storage-Material
- the capacitive design of a storage tank for long discharging cycles (24h - operation of the whole system)
- the accomodation of the HTST to the given conditions for charging (solar energy offer) and for discharging (energy demand of the linked process).

So the results of this study shall lead to predictions related to the previous three concerns.

Basic thermodynamical considerations

On the premises that the storage medium shall be the
Hybrid-material consisting of ceramic and salt and
that the heat transferring medium shall be a homogene
gas, for example air, the thermodynamical
considerations would have to start from the principle
of a three-dimensional triple-medium-storage-tank.
Supposing that the phase-change-material (PCM) is
distributed equally in the ceramic matrix, one can use
the calculation arrangements of a dual-medium-
storage-tank.
If it is furthermore presumed that there is no
temperature gradient inside the storage tank
orthogonal to the flow direction one can change into
the one-dimensional consideration.
This postulates an infinite coefficient of thermal
conductivity of the solid medium in horizontal
direction.
Later this inaccuracy will be reduced by a corrected
heat transfer coefficient between the solid and the
fluid medium.
Fig.8 shows the principle structure and designation of
a general one-dimensional dual-medium-storage-tank
/4/.
All appearing terms of the corresponding energy
balances are detailed in figure 9 and table 1.
The later occuring tables 2 and 3 also contain the
development of the system of partial differential

Energy balance for the solid medium:

$$\frac{\partial^2 E_s}{\partial x \cdot \partial \tau} = \frac{\partial \dot{Q}_{f/s}}{\partial x} - \frac{\partial \dot{Q}_{cs}}{\partial x} - \frac{\partial \dot{Q}_{amb}}{\partial x}$$

Partial differential of the energy stored ir the solid medium	Partial differential of the heat flux transferred between the solid and the fluid medium	Partial differential of the heat flux according to heat conduction in the solid medium	Partial differential of the ambient heat loss

Energy balance for the fluid medium:

$$\frac{\partial^2 E_f}{\partial x \cdot \partial \tau} + \frac{\partial \dot{H}_t}{\partial x} = -\frac{\partial \dot{Q}_{f/s}}{\partial x} - \frac{\partial \dot{Q}_{cf}}{\partial x}$$

Partial differential of the energy stored in the fluid medium	Partial differential of the total enthalpy of the fluid	Partial differential of the heat flux transferred between the solid and the fluid medium	Partial differential of the heat flux according to heat conduction in the fluid medium

Table 1: Energy balances for the solid and the fluid medium

equations, which describe the investigated thermodynamical model.

The energy balance for the solid medium says, that, concerning to the first law of thermodynamics, the changes of the inner energy stored is a consequence of the occuring heat transfer processes.

In the present case these are:

- the heat transfer between the solid and the fluid material
- the ambient heat loss
- the thermal conduction in vertical direction inside the solid medium.

The energy equation for the fluid first of all refers to a fixed differential element of the flowing medium. According to that the changes of the total enthalpy have to be taken into account besides the alterations of the energy stored and the heat flux from or to the solid material.

Tables 2 and 3 show the deduction of the partial differential equations for the inner energy of the solid and the fluid medium. The balances start from the exact principle of the one-dimensional dual-medium-storage-tank.

In order to ascertain the corresponding heat transfer characteristics with a reasonable time of calculating it is however necessary to use some simplifications during the modification of the equations.

Energy balance for the solid medium:

$$\frac{\partial^2 E_s}{\partial x \cdot \partial \tau} = \frac{\partial \dot{Q}_{f/s}}{\partial x} - \frac{\partial \dot{Q}_{cs}}{\partial x} - \frac{\partial \dot{Q}_{amb}}{\partial x}$$

$$E_s = \int_{0}^{x} \rho_s \cdot A_s \cdot u_s \cdot dx$$

$$\rho_s = \text{const.} \qquad A_s = \text{const.}$$

$$\frac{\partial^2 E_s}{\partial x \cdot \partial \tau} = \rho_s \cdot A_s \cdot \frac{\partial u_s}{\partial \tau}$$

$$\frac{\partial \dot{Q}_{f/s}}{\partial x} = \alpha \cdot U \cdot (t_f - t_s)$$

$$\dot{Q}_{cs} = -\lambda_s \cdot A_s \cdot \frac{\partial t_s}{\partial x}$$

$$\lambda_s = \text{const.}$$

$$\frac{\partial \dot{Q}_{cs}}{\partial x} = -\lambda_s \cdot A_s \cdot \frac{\partial^2 t_s}{\partial x^2}$$

$$\frac{\partial \dot{Q}_{amb}}{\partial x} = k_{amb} \cdot U_{amb} \cdot (t_s - t_{amb})$$

Differential equation for the inner energy of the solid medium:

$$\rho_s \cdot A_s \cdot \frac{\partial u_s}{\partial \tau} = \alpha \cdot U \cdot (t_f - t_s) + \lambda_s \cdot A_s \cdot \frac{\partial^2 t_s}{\partial x^2} - k_{amb} \cdot U_{amb} \cdot (t_s - t_{amb})$$

Table 2: Energy equation for the solid medium

<u>Energy balance for the fluid medium:</u>

$$\frac{\partial^2 E_f}{\partial x \cdot \partial \tau} \quad + \quad \frac{\partial \dot{H}_t}{\partial x} \quad = \quad -\frac{\partial \dot{Q}_{f/s}}{\partial x} \quad - \quad \frac{\partial \dot{Q}_{cf}}{\partial x}$$

$$\frac{\partial E_f}{\partial x} = (u_f + \frac{v_f^2}{2} + g \cdot z_f) \cdot \rho_f \cdot A_f$$

$$\rho_f = const. \qquad v_f = const. \qquad A_f = const. \qquad z_f = const.$$

$$\frac{\partial^2 E_f}{\partial x \cdot \partial \tau} = A_f \cdot \rho_f \cdot \frac{\partial u_f}{\partial x}$$

$$\dot{H}_t = \rho_f \cdot A_f \cdot v_f \cdot (u_f + \frac{p_f}{\rho_f} + \frac{v_f^2}{2} + g \cdot z_f)$$

$$\frac{\partial \dot{H}_t}{\partial x} = \rho_f \cdot A_f \cdot v_f \cdot \frac{\partial u_f}{\partial x}$$

$$\frac{\partial \dot{Q}_{f/s}}{\partial x} = \alpha \cdot U \cdot (t_f - t_s)$$

$$\frac{\partial \dot{Q}_{cf}}{\partial x} = -\lambda_f \cdot A_f \cdot \frac{\partial^2 t_f}{\partial x^2}$$

Transformation of the coordinates:

$$\varphi = arc\ tan\ (\frac{1}{v_f})$$

$$z = x \cdot cos\ \varphi + \tau \cdot sin\ \varphi \quad ; \quad y = -x \cdot sin\ \varphi + \tau \cdot cos\ \varphi$$

Differential equation for the inner energy of the fluid medium:

$$A_f \cdot \rho_f \cdot \frac{\partial u_f}{\partial z} = \alpha \cdot U \cdot (t_s - t_f) + \lambda_f \cdot A_f \cdot \frac{\partial^2 t_f}{\partial x^2}$$

Table 3: Energy equation for the fluid medium

In detail the following assumptions are made:

- the temperature dependence of the material data
  (ρ, c, λ) during a charging- or discharging-cycle is
  neglectible
- the changing of the fluid velocity in flow
  direction is neglectible
- the heat transferring circumferences as well
  inside the storage tank as at the outside border
  are constant
- the heat transfer coefficient is constant in flow
  direction.

The first supposition unfortunately was necessary as
there were no existing curves of the investigated
storage media for the examined high temperature range.
In future it will be however easily possible to
substitute the constant values in the program by
temperature curves achieved from planned measurements.
The second simplification is no crucial point, because
the pressure loss inside the flow canals will only be
in the range of 10^2Pa, so that the decrease of the gas
velocity will be very small over the whole flow path.
As the here investigated model presumes a cylindrical
vessel and cylindrical flow canals the third
assumption is exactly fulfilled.
Considering the fourth and last point one has to take
into account, that the heat transfer coefficient in
general is depending on the fluid velocity, on the

changes of the material data with the temperature and so on the temperature difference between the heat transferring media. As the maxima of these differences won't be very high during the examined cycles and as the first two suppositions have to be attended to, the inexactitude, that ensues from this assumption can be accepted.

All these presumptions lead to the definitive form of the partial differential equations for the inner energy of the solid and the fluid medium as they are used in the simulation program.

Table 4 shows these equations with the constant values A to E including all parameters depending on the geometry or the type of the storage media.

Thus the developed program offers the chance to involve different cases with changing specific values of the materials and with optional variable geometries into the same calculation routine.

Partial differential equation for the solid medium:

$$\frac{\partial u_s}{\partial \tau} = A \cdot (t_f - t_s) + B \cdot \frac{\partial^2 t_s}{\partial x^2} - C \cdot (t_s - t_{amb})$$

Partial differential equation for the fluid medium:

$$\frac{\partial u_f}{\partial z} = D \cdot (t_s - t_f) + E \cdot \frac{\partial^2 t_f}{\partial x^2}$$

Constants depending on the geometry and on the material data:

$$A = \frac{\alpha \cdot U}{\rho_s \cdot A_s}$$

$$B = \frac{\lambda_s}{\rho_s}$$

$$C = \frac{k_{amb} \cdot U_{amb}}{A_s \cdot \rho_s}$$

$$D = \frac{\alpha \cdot U}{\rho_f \cdot A_f}$$

$$E = \frac{\lambda_f}{\rho_f}$$

Table 4: Differential equations of the inner energy for a simplified one-dimensional dual-medium-storage-tank

Consideration of the phase-change

As there were no detailed informations available
related to the thermal characteristics of the
phase-change-materials (PCM), the shares of the latent
energy were considered in the energy balances as shown
in figure 10.

The diagram qualitatively shows the inner energy of
the PCM for different temperatures.

First the inner energy of the PCM increases because of
the sensible heat capacity with rising temperature
until a fictitious maximum melting-temperature is
reached.

If still more energy is added to the system, this
energy will be stored by changing phase at a constant
temperature.

After 100% of the PCM has a liquid condition the
continuing energy input again is stored as sensible
energy with increasing temperature.

In most cases not the whole amount of a PCM melts or
crystallizes at a fixed point but during a finite
temperature range.

So there is a maximum temperature for every medium,
where all the quantity of the examined material has
changed phase.

The used calculation model says, that the whole PCM
exactly melts at the above mentioned maximum
melting-temperature and that it crystallizes at a
minimum point, which is the temperature, where in

reality the last parts of the PCM become solid. So the calculated values will always have a safety margin, because their quantity of the latent energy stored will always be lower than the real values. Thermodynamically this model conception means, that the specific heat capacity of the PCM increases, when the maximum melting-temperature is reached and in fact the increase is corresponding to the heat of fusion of the investigated medium.

Characteristic material data

In order to use the developed calculation routine for
a storage tank model, which is as near as possible to
the later aspired system, together with DFVLR typical
material data were chosen to perform the planned
optimizations and to achieve results, that render
possible to make relevant predictions for the design
of the storage system.
As it is likely, that in the near future only the
technical application of phase-change-materials with a
fusion point in the temperature range up to 900 $^{\circ}$C
will be accomplished, the chosen salt (PCM) should be
a sodium-magnesium-fluoride and for the concerning
ceramic-material it was decided upon magnesia.
Table 5 contains the relevant data for the pure
ceramic material and for the Hybrid-medium combined of
the phase-changing salt and the magnesia ceramic.
The fusion point for the Hybrid will be between 780 $^{\circ}$C
and 880 $^{\circ}$C, so that the first value was taken as the
fictitious crystallization-temperature and the last
value as the maximum melting-temperature for the
model.
Furthermore the table shows that the combined material
not only is advantageous because of its ability to
change phase but also because of its about 13% higher
specific heat only related to sensible energy storing.
However it has to be admitted that the material data
for the Hybrid-medium are not yet totally corroberated

Ceramic - material (Magnesia):

- Density:..................................ρ = 2900 kg/m^3

- Specific heat:...........................c = 1100 J/(kg·K)

- Coefficient of thermal conductivity:......λ =   5 W/(m·K)

Hybrid - material (Sodium-Magnesium-Fluoride):

- Mass share:..............................45 % salt / 55 % MgO

- Heat of fusion:.........................Δh_f = 620 kJ/kg

- Melting-/Crystallization - temperature:..t_m = 830 $^\circ$C $\pm$ 50 $^\circ$C

- Density:..................................ρ = 2500 kg/m^3

- Specific heat:...........................c = 1250 J/(kg·K)

- Coefficient of thermal conductivity:......λ =   5 W/(m·K)

Table 5: Material data of the storage media

by experimences, so that the given values should only
be considered as a first estimation.

Especially the coefficient of thermal conductivity and
its alterations with changing temperature are loaded
up with high insecurity.

Nevertheless the results, achieved with these data,
can show relevant tendencies and can facilitate
decisions in future.

Moreover it will be easily possible, as described in
the former chapter, to substitute the given data in
the calculation routine by results of new
measurements.

Characteristic design data of the system

Besides the material data there were some more
establishments necessary to achieve evident results
for the investigated high-temperature-storage-system.
As the whole development yet was in the first stage,
there were hardly any restrictions for the design of
the whole system.
So we exemplified chose some reasonable data, which
are given in table 6.
The 24 hour-performance of the whole solar plant was
seperated into a maximum discharging period of 16
hours and conclusive a charging cycle of 8 hours.
The power output during discharging had to be at least
300 kW with a minimum outlet temperature of 750 $^{\circ}$C and
a pressure of 10 bar. The concerning inlet temperature
of the gas should be 600 $^{\circ}$C constantly.
For the charging period it was determined that the
inlet temperature of the unpressurized gas was 1000 $^{\circ}$C
constantly.
As described before the fusion-point and the
crystallization-temperature of the phase-change-medium
was fixed at 880 $^{\circ}$C and 780 $^{\circ}$C, respectively.
By using those fixations it is possible to deduce or
estimate some relevant marginal data for the storage
system.
Table 7 shows these values, which are at the same time
the basic input data for the simulation program.
First it is estimated, that there are about 46 tons or

```
┌──────────────────────────────────────────────────────────────────────┐
│                                                                        │
│  Discharging - cycle:                                                  │
│                                                                        │
│  - Minimum heat flux output:.....................Q̇_D  =   300 kW       │
│                                                                        │
│  - Maximum length of a period:....................τ_D  =    16 h       │
│                                                                        │
│  - Constant inlet-temperature:..................t_{Din} =  600 °C      │
│                                                                        │
│  - Minimum outlet-temperature:.................t_{Dout} = 750 °C       │
│                                                                        │
│  - Pressure:.......................................p_D = 10^6 Pa       │
│                                                                        │
└──────────────────────────────────────────────────────────────────────┘
┌──────────────────────────────────────────────────────────────────────┐
│                                                                        │
│  Charging - cycle:                                                     │
│                                                                        │
│  - Length of a period:.............................τ_C  =    8 h       │
│                                                                        │
│  - Constant inlet-temperature:...................t_{Cin} = 1000 °C     │
│                                                                        │
│  - Pressure:.......................................p_C = 10^5 Pa       │
│                                                                        │
└──────────────────────────────────────────────────────────────────────┘
┌──────────────────────────────────────────────────────────────────────┐
│                                                                        │
│  Change of phase:                                                      │
│                                                                        │
│  - Maximum melting-temperature:...............t_{m,max} = 880 °C       │
│                                                                        │
│  - Minimum crystallization-temperature:........t_{c,min} = 780 °C      │
│                                                                        │
└──────────────────────────────────────────────────────────────────────┘
```

Table 6: Design data of the system

```
┌─────────────────────────────────────────────────────────────────────┐
│                                                                       │
│                    D E S I G N    D A T A                             │
│                                                                       │
├─────────────────────────────────────────────────────────────────────┤
│                                                                       │
│      Storage capacity    ($t_{in}$=600°C, $t_{out}$=750°C):           │
│                                                                       │
│      $E_{ST}$ = $2.5 \cdot 10^{10}$ J ≅ 7 MWh                         │
│                                                                       │
├─────────────────────────────────────────────────────────────────────┤
```

Mass of the active storage part (HYBRID):....m_{ST} = $4.67 \cdot 10^4$ kg

Solid storage volume (HYBRID):..............V_{ST} =   18.7 m^3

Mass of the active storage part (CERAMIC):...m_{ST} = $1.12 \cdot 10^5$ kg

Solid storage volume (CERAMIC):.............V_{ST} =   38.7 m^3

Mass flow rate (charging-cycle):.............$\dot{m}_C$ =   2.5 kg/s

Average heat flux input (charging-cycle):.....$\dot{Q}_C$ =   857 kW

Dimensions of the storage tank (filling factor = 70 % HYBRID):

Total volume:...............................V_t =   26.7 m^3

Height:.....................................H_{ST} =   9.47 m

Diameter:...................................D_{ST} =   1.89 m

Table 7: Design data of the HTST for the simulation model

less than 19 m^3 of the Hybrid-medium necessary to
reach the given storage capacity of approximately
7 MWh, whereas the volume of the pure ceramic would be
more than twice as much.

Here it has to be mentioned, that the total storage
capacity is related to the minimum design temperature
of 600 $^\circ$C (discharging inlet temperature).

Furthermore it can be calculated that the average
energy flow input will be about 850 kW with a gas mass
flow of 2.5 kg/s.

Finally the dimensions of the storage tank for an
exemplified filling-factor of 70% and the
Hybrid-medium are given.

All these data of tables 6 and 7 will be the basis for
the results described in the last chapter of this
report.

Way of optimizing

The optimizing of the whole storage system has to be carried out essentially under the following aspects:

1) Thermodynamically:

   -maximum heat transfer by

   - high heat transfer coefficients
   - large heat transfer surfaces

2) Technically:

   - high mechanical stability of the stonework by
   - limitation of the ratio storage height to storage diameter to values around five /2/

3) Economically:

   - investments as low as possible by
   - utilization of the maximum filling-factor (ratio of solid volume to total volume), as the main part of the investments will be determined by the size of the pressure vessel.

For a given cross-section area the two thermodynamical demands are partly contrarily, because the heat transferring surface increases with decreasing inner diameters of the gas flow canals, whereas the heat transfer coefficient has its maximum at the smallest diameter, which is necessary for a turbulent flow. Figure 11 shows the heat transfer coefficient as a function of the inner diameter of the flow canals and for different filling factors.

One can see that the turn from the turbulent to the laminar flow causes a considerable decrease of the heat transfer coefficient.

The concerning mathematical connexion is given in table 8.

The table starts with the deduction of the critical inner diameter of the flow canals, i. e. the smallest diameter which allows turbulent flow. As the thermodynamical model is based on cylindrical flow canals inside the storage tank, the critical Reynolds-Number will be in the range of 2300.

This relation then is used to get the above mentioned connexion between the heat transfer coefficient and the inner diameter of the flow canal d_i, which now substitutes the former variable d_{krit}.

Subsequently the heat transfer coefficient α is combined with the heat transferring circumference U.

The product $(\alpha \cdot U)$ was chosen, because - for the examined model - the circumference is proportional to the heat transferring surface (proportional factor: storage height). The concerning figure 12 shows the same tendencies as the diagram in figure 11, but one can see, that the heat transfer product $(\alpha \cdot U)$ for very small diameters d_i could even be higher than the product at the critical diameter.

This fact however won't be profitable in practice, because very small canal diameters would also cause

Critical diameter of the flow canals

$$Re_{crit} = \frac{w \cdot d_{crit}}{\upsilon} \qquad \Longrightarrow \qquad d_{crit} = \frac{\upsilon \cdot Re_{crit}}{w}$$

$$\dot{m} = \rho \cdot w \cdot A_{Qf} \qquad \Longrightarrow \qquad w = \frac{\dot{m}}{\rho \cdot A_{Qf}}$$

$$H/D =_{def.} 5$$
$$V_s = FF \cdot V_t = FF \cdot \frac{\pi}{4} \cdot D^2 \cdot H \qquad \Big\} \quad D = \left(\frac{4 \cdot V_s}{5 \cdot \pi \cdot FF} \right)^{1/3}$$

$$A_{Qf} = (1 - FF) \cdot A_t = (1 - FF) \cdot \frac{\pi}{4} \cdot D^2$$
$$A_{Qf} = (1 - FF) \cdot \frac{\pi}{4} \cdot \left(\frac{4 \cdot V_s}{5 \cdot \pi \cdot FF} \right)^{2/3}$$

$$w = \frac{\dot{m}}{\rho \cdot (1 - FF) \cdot \frac{\pi}{4} \cdot \left(\frac{4 \cdot V_s}{5 \cdot \pi \cdot FF} \right)^{2/3}}$$

$$d_{crit} = \frac{Re_{crit} \cdot \upsilon \cdot \rho \cdot (1-FF) \cdot \frac{\pi}{4} \cdot \left(\frac{4 \cdot V_s}{5 \cdot \pi \cdot FF} \right)^{2/3}}{\dot{m}}$$

Table 8: Heat transfer parameter

Heat transfer coefficient
$$\alpha = \frac{\lambda \cdot \text{Nu}}{d_i}$$
Turbulent: $\text{Nu} = 0.024 \cdot \text{Re}^{0.8} \cdot \text{Pr}^{1/3}$ [Laminar: $\text{Nu} = 1.86 \cdot (\text{Re} \cdot \text{Pr})^{0.33} \cdot (\frac{d_i}{H})^{0.33}$]   /5/
$$\text{Re} = \frac{\dot{m} \cdot d_i}{\upsilon \cdot \rho \cdot (1 - FF) \cdot \frac{\pi}{4} \cdot (\frac{4 \cdot V_s}{5 \cdot \pi \cdot FF})^{2/3}}$$
$$\alpha = \frac{\lambda \cdot 0.024 \cdot \text{Pr}^{0.33}}{d_i} \cdot \left\{ \frac{\dot{m} \cdot d_i}{\frac{\pi}{4} \cdot \upsilon \cdot \rho \cdot (1-FF)} \right\}^{0.8} \cdot \left\{ \frac{5 \cdot \pi \cdot FF}{4 \cdot V_s} \right\}^{0.533}$$

Heat transferring circumference
$$U = \pi \cdot d_i \cdot N_c$$ $$N_c = \frac{A_f}{\frac{\pi}{4} \cdot d_i^2} \qquad \Longrightarrow \qquad U = \frac{4 \cdot A_f}{d_i}$$
$$A_f = (1 - FF) \cdot \frac{\pi}{4} \cdot D^2 = (1 - FF) \cdot \frac{\pi}{4} \cdot (\frac{4 \cdot V_s}{5 \cdot \pi \cdot FF})^{2/3}$$
$$U = \frac{\pi \cdot (1 - FF)}{d_i} \cdot (\frac{4 \cdot V_s}{5 \cdot \pi \cdot FF})^{2/3}$$

Product $(\alpha \cdot U)$
$$\alpha \cdot U = \frac{\pi \cdot (1-FF) \cdot \lambda \cdot 0.024 \cdot \text{Pr}^{0.33}}{d_i^2} \cdot (\frac{4 \cdot \dot{m} \cdot d_i}{\pi \cdot \upsilon \cdot \rho \cdot (1-FF)})^{0.8} \cdot (\frac{4 \cdot V_s}{5 \cdot \pi \cdot FF})^{0.133}$$

Table 8 (continuation): Heat transfer parameter

very small webs between them, so that the mechanical stability of the relining would not longer be given.

The four resulting equations for d_{krit}, α, U and $(\alpha \cdot U)$ were deducted as general as possible, so that the examined variables are only depending on the corresponding material data and the nominal values of the solid storage volume V_s, the filling factor FF and the mass flow rate of the gas $\dot{m}$.

In addition to that it has to be noticed that during the calculation of the simulation program the heat transfer coefficient is reduced as follows:

$$\alpha_{red} = \frac{\alpha \cdot \dfrac{\lambda}{\phi \cdot s}}{\alpha + \dfrac{\lambda}{\phi \cdot s}} \qquad \text{with} \qquad \phi = \frac{1}{6} \qquad /6/.$$

This reduction was necessary to adjust the wrong assumption, that the coefficient of the thermal conductivity of the solid medium is infinite perpendicular to the flow direction.

Results

Figure 13 shows the temperature-profiles versus the storage tank height for a 24 hour-cycle. The curves are plotted with a time difference of 4 hours.

The initial profile at 0h was taken from a programm run for a 72 hour performance (3 charging- and 3 discharging-cycles alternating), so that it can be looked upon as the typical temperature profile for an "empty" storage tank.

The fact that the 20h- and the nearly coinciding 24h-curve are even removed to lower storage temperatures as the initial profile is caused by the reason, that the calculation was not stopped when the outlet-temperature fell below the minimum value of 750 $^\circ$C.

The diagram shows, that after 4 hours of charging the area, which contains latent energy stored, has climbed down to 5.5 m.

However it has to be attended to the fact, that from a level of 7.5 m up to the top of the tank there is already sensible energy stored with temperatures, that are about 100 $^\circ$C higher than the maximum melting-temperature of the PCM.

Looking at the 8h-profile, which represents the condition of the storage unit at the end of the charging-cycle, one can determine, that the part of the tank, which uses the phase-change of the incorporated salt, is more than 80%.

Considering the beginning of the discharging-cycle it is obvious, that unfortunately mainly the sensible energy stored in the lower areas is transferred to the gas flow, so that the 12h - profile shows a temperature difference of about 200 $^{\circ}$C between top and bottom of the tank.

Finally the 20h - and analogously the not plotted end-profile of the discharging-period indicate, that the investigated concept of a HTST will probably only be convenient for long time storing as it is examined in this study, because during the last part of the discharging-cycle the latent energy is utilized in a high degree.

This point of view can be emphasized by the following two diagrams of figure 14a and 14b, where the share of the melted PCM is drawn versus the storage tank height.

As the height is directly proportional to the storage volume one can imagine, that the area above every curve represents the quantity of the liquified salt.
So it is evident, that at the end of the charging period (fig. 14a, 8h-profile) about half of the incorporated PCM has stored latent energy.

Furthermore one can conclude, that it will be reasonable, according to the given conditions of the storage system, to use the Hybrid-material only from a determined level up to the top of the tank.

The lower part of the unit then could be filled with

conventional solid material.

The analogous diagram for the discharging-cycle
(fig.14b) illustrates the above mentioned inference,
that the use of the Hybrid-medium will only be
efficient for cycles with long periods.

Especially the profile after 4 hours of discharging,
which nearly coincides with the discharging initial
curve (0h) shows, that only a very small part of the
latent energy stored could be used.

In contrast to this the distances between the later
profiles make it obviously, that the efficiency of the
storage of latent energy increases with proceeding
time.

The results of the direct comparison between the
Hybrid- and the pure ceramic-material are illustrated
in figure 15. There the transferred heat flux related
to a m^3 of storage volume is shown versus the time of
a complete charging- and discharging-period. The
energy flux density of the Hybrid-storage-model during
discharging is around twice as high as the output of
the ceramic-model.

Moreover the opposition of the maximum values of the
heat flux density, 42 kW/m^3 (Hybrid) opposite to
18 kW/m^3 (ceramic), corroborate the capacitive
advantages of the combined material.

Figure 16 contains the results of a variation of the
mass flow during a discharging cycle.

The relining was Hybrid with a filling factor of 70%.
For two different mass flow rates, namely 1.2 kg/s and
1.8 kg/s, the time-slopes of the outlet-temperature
and the energy flux output is represented.
The curves shows, that it is possible to observe
different limited discharging-outlet-temperatures,
which can be necessary for the combined energy
conversion system, by varying only the gas mass flow.
For example, if the minimum permitted outlet-
temperature is 750 $^\circ$C, an average heat flux output of
about 300 kW is possible during the 16 hours of
discharging with a mass flow of 1.2 kg/s.
Raising the mass flow to 1.8 kg/s the concerning
energy output increases to around 350 kW on the
average, whereas the minimum outlet temperature is
about 700 $^\circ$C.
The last diagram, figure 17, contains another
important valuation parameter for a storage system,
namely the total efficiency ε_t. Opposite to some other
declarations in literature ε_t here is defined as the
ratio of the effective energy flux output to the whole
amount of the concerning input. So even if the energy
stored could be discharged completely, the total
efficiency couldn't reach the 100% mark because of the
losses of the preceding charging-cycle.
Considering this definition of ε_t it becomes evident,
that the achieved values of about 60% point to the
fact, that the investigated storage model utilizes the
energies being at disposal very well.

Conclusion

By developing and using a simulation program for the
performance characteristic of a High-Temperature-
Storage-Tank this study rendered possible to make
relevant declarations for the fundamental design of a
HTST and for the basic optimization of important
parameters.

The essential inference that can be drawn from the
obtained findings is, that a HTST using a Hybrid-
storage-medium for sensible and latent heat will be
efficient in combination with solar power plants,
because its capacitive advantages especially occur, if
the storage unit has to operate in long cycles,
exemplified in the range of 24 hours.

The investigated thermodynamical model furthermore
shows, that, according to the existing conditions of
the linked energy source and conversion system, it
will be reasonable to use the combined medium only in
the upper parts of the storage unit and to fill the
rest of the tank with conventional ceramic material.
Consequently it can finally be ascertained, that, if
the technical and chemical problems referring to the
Hybrid-storage-medium can be solved in future, the
continuation of the development of a High-Temperature-
Storage-Tank using this combined material is of great
promise for many applications of thermal energy
storage.

FIGURES

Schematic System Structure

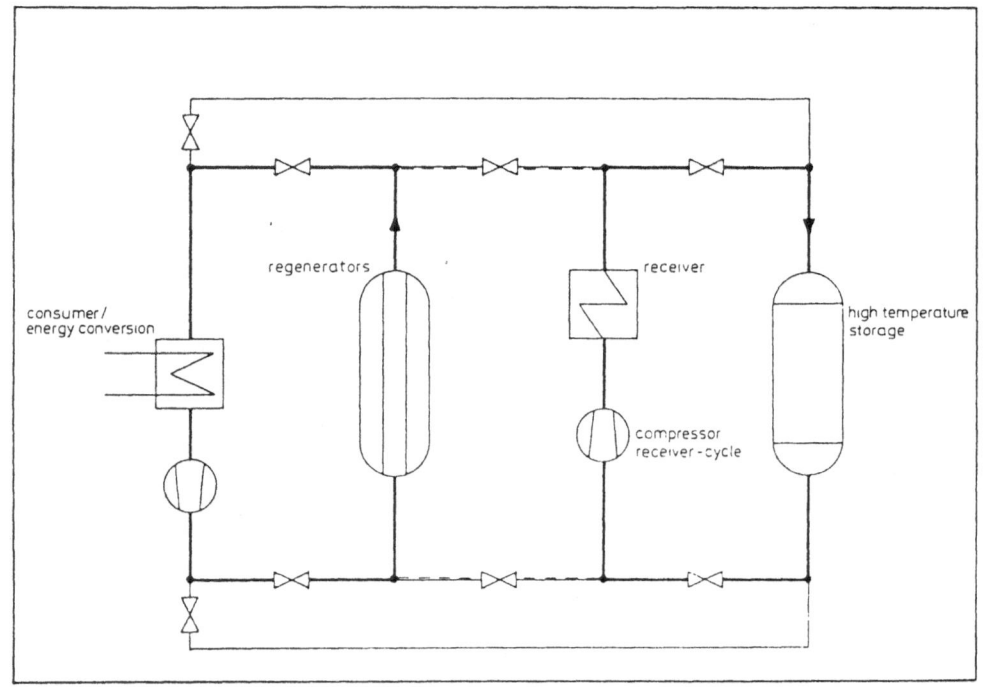

Fig. 1   Charging of the storage

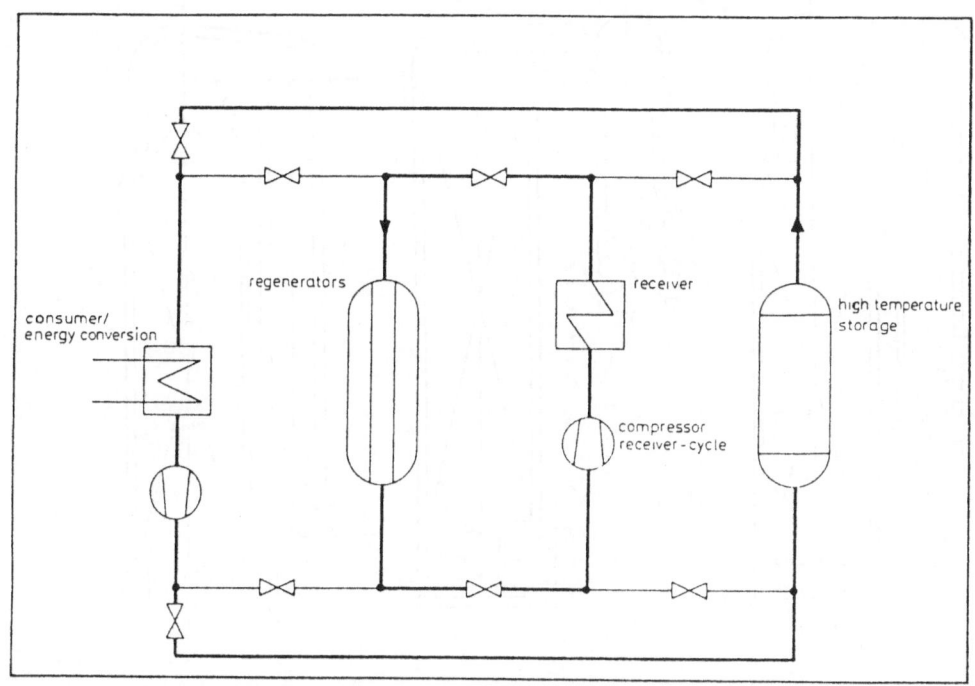

Fig. 2   Discharging of the storage

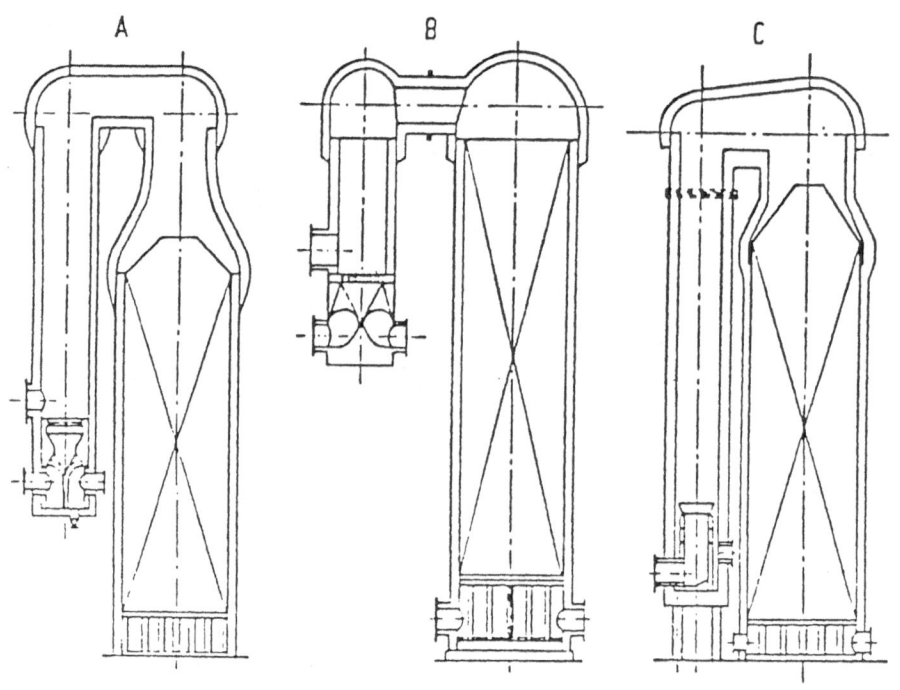

Fig. 3: HIGH-TEMPERATURE BLAST STOVE WITH OUTSIDE COMBUSTION CHAMBER /1/

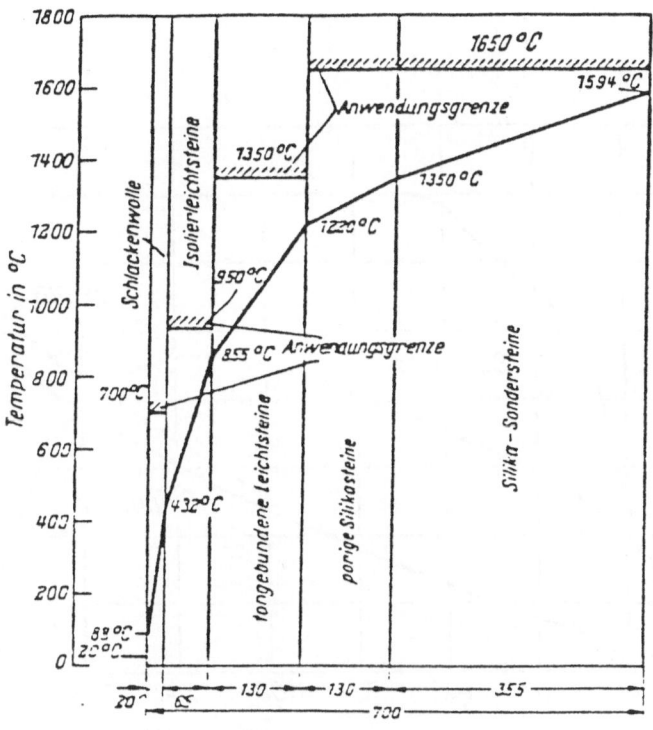

Fig. 4: TEMPERATURE-CURVE IN THE STONEWORK OF THE DOME /2/

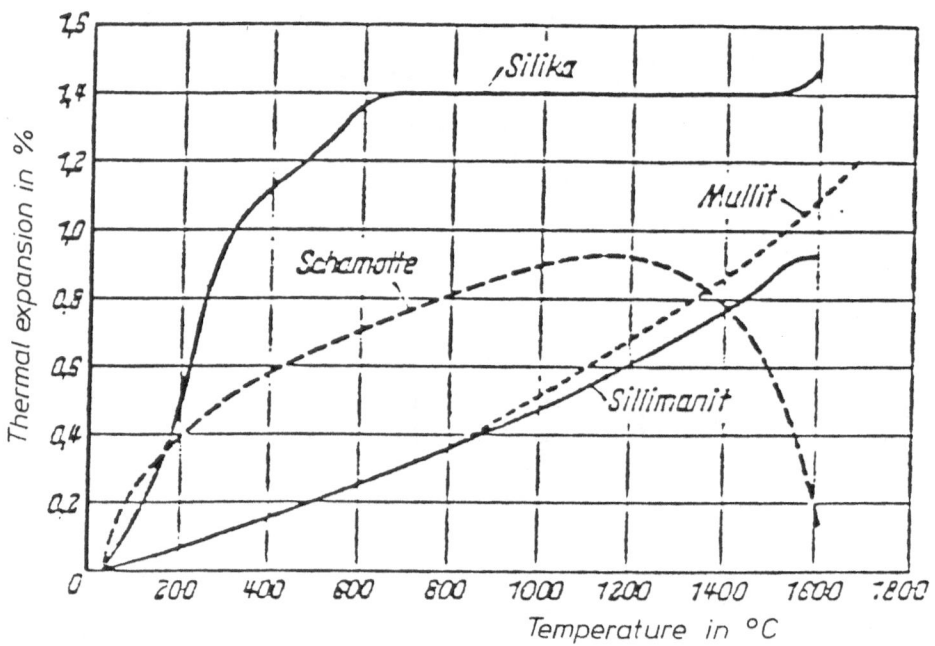

Fig. 5   Thermal expansion of different stonework materials   /2/

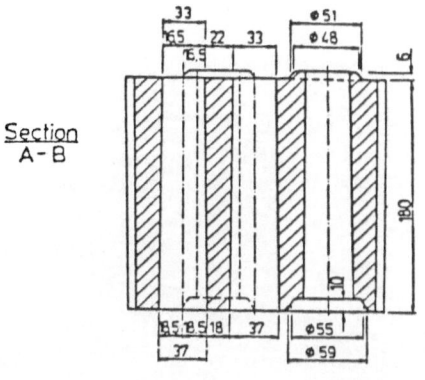

Section
A – B

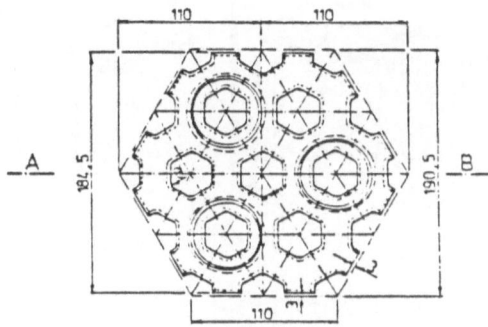

Fig. 6: RELINING FORM FOR A HOT BLAST STOVE /3/

Fig. 7: RELINING FORMS FOR A HIGH-TEMPERATURE-STORAGE-TANK /1/

REFERENCES

/1/   Karl Engel, Dieter Winzer:
      "Bauweise und Betriebsergebnisse von
      Hochtemperaturwinderhitzern",
      Stahl und Eisen 95 (1975), Nr. 17

/2/   Erich E. Hofmann:
      "Moeglichkeiten fuer den Bau von Winderhitzern
      fuer hohe Windtemperaturen",
      Stahl und Eisen 83 (1963), Nr. 4

/3/   SSPS TECHNICAL REPORT No. 2/86
      Proceedings of the 1st IEA-SSPS Task IV Status
      Meeting on "High Temperature Thermal Storage",
      Tabernas / Spain,   1986

/4/   Walter Bitterlich:
      "Speicher fuer thermische Energie",
      Energie- und Kraftwerkstechnik,
      Universitaet Gesamthochschule Essen,
      Dezember 1985 (not published)

/5/   Fritz Steimle:
      "Waermeuebertragung I", Formelverzeichnis,
      Thermodynamik und Klimatechnik,
      Universitaet Gesamthochschule Essen,
      1981 (not published)

/6/   Helmuth Hausen:
      "Waermeuebertragung im Gegenstrom, Gleichstrom
      und Kreuzstrom",
      Springer Verlag
      Berlin Heidelberg New York   1976

APPENDIX

Numerical output of the simulation program

The following numerical outputs of the simulation
program contain the data, which were used to make the
diagrams of figures 13 to 17.

Independent from the variation of the parameters every
storage cycle was calculated for a period of 72 hours
(3 charging- and 3 discharging-cycles alternating), in
order to achieve a stabilized performance
characteristic.

As, however, only the last 24h-cycles are important
for the accomplished assessments, only those data are
added here.

So the storage periods concerning to the above
mentioned figures begin at 48h, which corresponds to
0h in the diagrams, and end at 72h, which corresponds
to 24h in the diagrams.

The different results are specified by Roman numbers
and the concerning varied parameters are declared on
every first page of the output.

NOMENCLATURE OF THE VARIABLES OF THE NUMERIC OUTPUT

(Same sequence as in the output)

AB	
AE	
B	
C	Constants of the partial
DB	differential equations
DE	
EB	
EE	
EST	maximum heat capacity of the storage tank
EMST	mass of the solid storage medium
VS	volume of the solid storage medium
QPCHA	control value
RMPGB	mass flow (charging)
RMPGE	mass flow (discharging)
H	storage tank height
DA	storage tank diameter
DKRIT	critical diameter of the flow canals
DI	inner diameter of the flow canals
RET	turbulent Reynolds-Number
REL	laminar Reynolds-Number
RNUT	turbulent Nusselt-Number
RNUL	laminar Nusselt-Number
ALPHAT	turbulent heat transfer coefficient
ALPHAL	laminar heat transfer coefficient
ALPHAB	reduced turbulent heat transfer coeff.
ALPHAE	reduced laminar heat transfer coeff.
NBOHR	number of flow canals
AG	cross-sectional area for the fluid medium
AK	cross-sectional area for the solid medium
U	heat transferring circumference
S	web-width between the flow canals

===

X	flow path (storage tank height)
TK	temperature of the solid medium
TG	temperature of the fluid medium
UK	inner energy of the solid medium
UG	inner energy of the fluid medium
ESP	energy stored
SA	mass share of melted salt
QK	
QP	control values for
QG	the energy balances
(QG/TAUL)	

- 262 -

NUMERICAL OUTPUT   No. I

```
     II       =M NUMBER OF CALCULATION POINTS IN X-DIRECTION
   1000.00000  =TGEBE [C] INLET TEMPERATURE FLUID (CHARGING)
    600.000000  =TGEEN [C] INLET TEMPERATURE FLUID (DISCHARGING)
    880.000000  =TKS [C] MELTING POINT
    780.000000  =TKK [C] CRYSTALLIZATION POINT
     20.0000000  =TU [C] AMBIENT TEMPERATUR
   2500.00000  =RHOK [KG/M**3] DENSITY SOLID MEDIUM
   1100.00000  =CKK [J/(KG*K)] SPECIFIC HEAT CERAMIC
   1433.33007  =CKS [J/(KG*K)] SPECIFIC HEAT PHASE-CHANGE-MATERIAL
      5.00000000  =RLAMK [W/(M*K)] COEFF. OF THERMAL CONDUCT. SOLID MED.
   620000.000  =RK [J/KG] HEAT OF FUSION
      0.449999988  =RMS MASS-SHARE OF SALT
      0.310500025  =RHOGB [KG/M**3] DENSITY FLUID (CHARGING)
      3.66800022  =RHOGE [KG/M**3] DENSITY FLUID (DISCHARGING)
   1163.00000  =CGB [J/(KG*K)] SPECIFIC HEAT FLUID (CHARGING)
   1132.00000  =CGE [J/(KG*K)] SPECIFIC HEAT FLUID (DISCHARGING)
      0.727999806E-01 =RLAMGB [W/(M*K)] COEFF. OF THERMAL COND. FLUID CHARG.
      0.665000081E-01 =RLAMGE [W/(M*K)] COEFF. OF THERM. COND. FLUID DISCH.
      0.709999978  =PRGB PRANDTL-NUMBER (CHARGING)
     69.0000000  =PRGE PRANDTL-NUMBER (DISCHARGING)
    800.000000  =TKVA NOMINAL TEMPERATURE FOR HEAT LOSS
      1.00000000  =DTKVU TEMPERATURE GRADIENT FOR HEAT LOSS
   300000.000  =QPDIS[W] ENERGY OUTPUT (DISCHARGING)
    57600.0000  =TAUDIS [S] DISCHARGING-TIME
    28800.0000  =TAUCHA [S] CHARGING-TIME
      0.699999988  =ETASTO EFFICIENCY OF STORING (STARTVALUE)
    700.000000  =TCOUT [C] OUTLET-TEMPERATURE FLUID (CHARGING)
    750.000000  =TDOUT [C] OUTLET-TEMPERATURE FLUID (DISCHARGING)
      0.600000023  =FF FILLING FACTOR
      0.143700002E-03 =RNUEB [M**2/S] VISCOSITY FLUID (CHARGING)
      0.110500003E-02 =RNUEE [M**2/S] VISCOSITY FLUID (DISCHARGING)
    200.000000  =DELTAT [C] AVERAGE TEMPERATURE DIFFERENCE CERAMIC
```

```
AB=  0.298863828
AE=  0.907190442E-01
B=  0.199999986E-02
C=  0.118000009E-07
DB=   3608.11132
DE=   92.7123107
EB=  0.234460413
EE=  0.181289697E-01
EST   0.246857113E+11
EMST   46664.8984
VS    18.6659545
QPCHA  -0.230789184E-03
RMPGB   2.50000000
RMPGE   1.79999923
H    9.96742248
DA    1.99348449
DKRIT  0.516941249E-01
DI  0.519999973E-01
RET    2333.72534
REL    18.4973144
RNUT    10.6063575
RNUL    3.47718334
ALPHAT    14.8488960
ALPHAL    4.44678211
ALPHAB    14.5641431
ALPHAE    4.42089557
NEOHR        588
AG    1.24874496
AK    1.87241458
U    96.0572967
S   0.394999422E-01
```

```
TAU=    3600.7   TOUT=   709.60   SUMMET= 0.245924E+07
TAU=    7201.4   TOUT=   755.97   SUMMET= 0.508749E+07
TAU=   10802.5   TOUT=   796.35   SUMMET= 0.787269E+07
TAU=   14403.2   TOUT=   829.08   SUMMET= 0.107901E+08
TAU=   18003.9   TOUT=   853.63   SUMMET= 0.138098E+08
TAU=   21604.6   TOUT=   870.29   SUMMET= 0.168997E+08
TAU=   25205.3   TOUT=   880.99   SUMMET= 0.198810E+08
TAU=   28505.9   TOUT=   885.79   SUMMET= 0.226140E+08
*******************************************************
DISCHARGING   TAU=   28800.
*******************************************************
TAU=   32419.5   TOUT=   881.47   SUMMET= 0.324546E+07
TAU=   36041.4   TOUT=   853.74   SUMMET= 0.638659E+07
TAU=   39663.4   TOUT=   828.72   SUMMET= 0.943170E+07
TAU=   43285.3   TOUT=   807.15   SUMMET= 0.123924E+08
TAU=   46907.3   TOUT=   789.16   SUMMET= 0.152821E+08
TAU=   50529.2   TOUT=   772.65   SUMMET= 0.181037E+08
TAU=   54151.1   TOUT=   759.74   SUMMET= 0.208668E+08
TAU=   57471.3   TOUT=   749.37   SUMMET= 0.233615E+08
TAU=   61093.2   TOUT=   741.15   SUMMET= 0.260473E+08
TAU=   64715.1   TOUT=   736.20   SUMMET= 0.287096E+08
TAU=   65017.0   TOUT=   735.89   SUMMET= 0.289304E+08
TAU=   68638.9   TOUT=   733.25   SUMMET= 0.315808E+08
TAU=   71959.0   TOUT=   729.59   SUMMET= 0.339984E+08
TAU=   75581.0   TOUT=   726.01   SUMMET= 0.366259E+08
TAU=   79202.9   TOUT=   723.07   SUMMET= 0.392347E+08
TAU=   82824.9   TOUT=   718.36   SUMMET= 0.418358E+08
TAU=   86145.0   TOUT=   714.20   SUMMET= 0.442025E+08
*******************************************************
CHARGING     TAU=   86401.
*******************************************************
TAU=   90001.7   TOUT=   671.59   SUMMET= 0.232995E+07
TAU=   93602.4   TOUT=   719.53   SUMMET= 0.482284E+07
TAU=   97203.2   TOUT=   764.43   SUMMET= 0.748442E+07
TAU=  100803.9   TOUT=   802.86   SUMMET= 0.102969E+08
TAU=  104404.6   TOUT=   833.06   SUMMET= 0.132330E+08
TAU=  108005.3   TOUT=   855.16   SUMMET= 0.162634E+08
TAU=  111606.0   TOUT=   869.88   SUMMET= 0.192604E+08
TAU=  114906.6   TOUT=   878.60   SUMMET= 0.219933E+08
*******************************************************
DISCHARGING   TAU=   115200.
*******************************************************
TAU=  118819.6   TOUT=   869.50   SUMMET= 0.320007E+07
TAU=  122441.5   TOUT=   842.81   SUMMET= 0.629973E+07
TAU=  126063.4   TOUT=   818.17   SUMMET= 0.930629E+07
TAU=  129685.4   TOUT=   796.91   SUMMET= 0.122297E+08
TAU=  133307.3   TOUT=   779.56   SUMMET= 0.150825E+08
TAU=  136929.3   TOUT=   764.30   SUMMET= 0.178732E+08
TAU=  140551.2   TOUT=   751.83   SUMMET= 0.206061E+08
TAU=  143871.3   TOUT=   742.91   SUMMET= 0.230761E+08
TAU=  147493.3   TOUT=   734.50   SUMMET= 0.257385E+08
TAU=  151115.2   TOUT=   729.40   SUMMET= 0.283774E+08
TAU=  154737.2   TOUT=   726.08   SUMMET= 0.310050E+08
TAU=  158359.1   TOUT=   722.77   SUMMET= 0.336145E+08
TAU=  161981.1   TOUT=   717.69   SUMMET= 0.362122E+08
TAU=  165603.0   TOUT=   713.48   SUMMET= 0.387941E+08
TAU=  169225.0   TOUT=   709.82   SUMMET= 0.413570E+08
TAU=  172545.1   TOUT=   703.90   SUMMET= 0.436930E+08
*******************************************************
CHARGING     TAU=   172801.
*******************************************************
TAU=  176401.8   TOUT=   663.50   SUMMET= 0.230735E+07
TAU=  180002.5   TOUT=   709.72   SUMMET= 0.476732E+07
TAU=  183603.1   TOUT=   754.72   SUMMET= 0.739349E+07
TAU=  187203.8   TOUT=   794.29   SUMMET= 0.101725E+08
TAU=  190804.5   TOUT=   826.00   SUMMET= 0.130807E+08
TAU=  194405.2   TOUT=   849.94   SUMMET= 0.160888E+08
TAU=  198005.9   TOUT=   866.21   SUMMET= 0.190864E+08
TAU=  201306.5   TOUT=   876.26   SUMMET= 0.218193E+08
*******************************************************
DISCHARGING   TAU=   201600.
*******************************************************
TAU=  205219.5   TOUT=   866.60   SUMMET= 0.318888E+07
TAU=  208841.4   TOUT=   840.22   SUMMET= 0.627860E+07
TAU=  212463.4   TOUT=   815.77   SUMMET= 0.927612E+07
TAU=  216085.3   TOUT=   794.50   SUMMET= 0.121909E+08
TAU=  219707.3   TOUT=   777.31   SUMMET= 0.150355E+08
TAU=  223329.2   TOUT=   761.90   SUMMET= 0.178179E+08
TAU=  226951.1   TOUT=   750.24   SUMMET= 0.205449E+08
TAU=  230573.1   TOUT=   739.78   SUMMET= 0.232299E+08
TAU=  234195.0   TOUT=   732.40   SUMMET= 0.258847E+08
TAU=  241137.1   TOUT=   724.71   SUMMET= 0.309207E+08
TAU=  241438.9   TOUT=   724.23   SUMMET= 0.311377E+08
TAU=  245060.9   TOUT=   719.02   SUMMET= 0.337418E+08
TAU=  248682.8   TOUT=   714.68   SUMMET= 0.363236E+08
TAU=  252304.8   TOUT=   711.06   SUMMET= 0.388938E+08
TAU=  255926.7   TOUT=   706.19   SUMMET= 0.414516E+08
TAU=  258945.0   TOUT=   701.32   SUMMET= 0.435650E+08
```

```
        48.00 TAUH IN H.MIN

    X       TK        TG        UK          UG         ESP          SA
  0.00   780.000   703.488  1129146.00   796348.75  0.406785E+11  0.55250
  1.00   780.000   685.749  1079838.00   776268.44  0.371430E+11  0.37577
  1.99   780.000   663.896  1015589.69   751530.31  0.335809E+11  0.14549
  2.99   747.588   640.730   934483.44   725306.75  0.299848E+11  0.00000
  3.99   682.209   623.530   852760.25   705836.25  0.263430E+11  0.00000
  4.98   644.484   614.296   805604.56   695383.50  0.226295E+11  0.00000
  5.98   628.983   609.093   786227.81   689493.06  0.187607E+11  0.00000
  6.98   619.864   605.537   774828.50   685468.37  0.145912E+11  0.00000
  7.97   613.313   602.975   766641.00   682567.31  0.100419E+11  0.00000
  8.97   608.209   601.169   760260.06   680523.31  0.515344E+10  0.00000
  9.97   604.212   600.000   755264.75   679200.00  0.000000E+00  0.00000
QK-0.598905159E+11
QP,QG,(QG/TAUL)  210.867492        19176960.0        333.077880


****************************************************************
CHARGING    TAU=  172800.
****************************************************************


        49.00 TAUH IN H.MIN

    X       TK        TG         UK          UG         ESP          SA
  0.00   880.000  1000.000  1292149.00  1163000.00  0.444176E+11  0.68871
  1.00   872.584   952.144  1195569.00  1107343.00  0.386139E+11  0.37577
  1.99   847.423   916.498  1099869.00  1065887.00  0.332588E+11  0.14549
  2.99   812.553   883.032  1015469.81  1026966.25  0.283234E+11  0.00000
  3.99   761.970   845.983   952461.62   983878.81  0.237319E+11  0.00000
  4.98   721.107   805.577   901383.06   936885.69  0.194071E+11  0.00000
  5.98   692.070   767.281   865086.44   892347.69  0.152861E+11  0.00000
  6.98   669.823   733.885   837277.62   853508.25  0.113147E+11  0.00000
  7.97   652.414   705.735   815515.94   820769.56  0.745890E+10  0.00000
  8.97   638.713   682.452   798390.37   793691.87  0.369384E+10  0.00000
  9.97   627.987   663.457   784983.31   771601.06  0.000000E+00  0.00000
QK 0.444175810E+11
QP,QG,(QG/TAUL) -978.497314        -3731077.00       -1037.31396


        50.00 TAUH IN H.MIN

    X       TK        TG         UK          UG         ESP          SA
  0.00   909.433  1000.000  1415790.00  1163000.00  0.476996E+11  1.00000
  1.00   880.000   959.270  1276108.00  1115631.00  0.414195E+11  0.63122
  1.99   880.000   928.605  1157455.00  1079968.00  0.357421E+11  0.20594
  2.99   862.111   906.343  1077637.00  1054077.00  0.305279E+11  0.00000
  3.99   823.453   881.753  1029315.31  1025478.75  0.256126E+11  0.00000
  4.98   786.277   852.011   982844.75   990889.50  0.209184E+11  0.00000
  5.98   753.864   820.305   942328.87   954015.44  0.164273E+11  0.00000
  6.98   725.647   789.152   907058.37   917784.06  0.121128E+11  0.00000
  7.97   701.393   759.882   876746.25   883742.94  0.795139E+10  0.00000
  8.97   680.828   733.299   851033.69   852826.94  0.393067E+10  0.00000
  9.97   663.658   709.655   829571.06   825328.75  0.000000E+00  0.00000
QK 0.476996239E+11
QP,QG,(QG/TAUL) -844.178222        -7001645.00       -979.033691
```

```
      51.00 TAUH IN H.MIN

    X       TK        TG         UK          UG          ESP         SA
 0.00    961.461  1000.000  1480825.00  1163000.00  0.504999E+11  1.00000
 1.00    880.000   969.333  1368011.00  1127335.00  0.438537E+11  0.96062
 1.99    880.000   934.776  1213510.00  1087145.00  0.378311E+11  0.40685
 2.99    880.000   913.587  1115886.00  1062502.00  0.323968E+11  0.05695
 3.99    863.275   897.358  1079092.00  1043627.06  0.272761E+11  0.00000
 4.98    833.956   878.503  1042443.06  1021699.12  0.223268E+11  0.00000
 5.98    804.767   855.623  1005957.69   995089.50  0.175481E+11  0.00000
 6.98    776.699   830.520   970872.31   965894.44  0.129364E+11  0.00000
 7.97    750.489   804.628   938109.62   935782.00  0.848294E+10  0.00000
 8.97    726.589   779.071   908235.44   906059.19  0.417562E+10  0.00000
 9.97    705.326   754.625   881656.19   877628.94  0.000000E+00  0.00000
QK 0.504999485E+11
QP,QG,(QG/TAUL) -713.427490      -9789680.00      -914.977294


      52.00 TAUH IN H.MIN

    X       TK        TG         UK          UG          ESP         SA
 0.00    983.448  1000.000  1508309.00  1163000.00  0.528555E+11  1.00000
 1.00    932.680   983.776  1444849.00  1144132.00  0.459659E+11  1.00000
 1.99    880.000   953.818  1282811.00  1109291.00  0.396023E+11  0.65524
 2.99    880.000   925.264  1158226.00  1076082.00  0.339075E+11  0.20870
 3.99    880.000   907.754  1109555.00  1055718.00  0.286170E+11  0.03425
 4.98    865.518   894.215  1081896.00  1039972.69  0.235046E+11  0.00000
 5.98    842.351   878.631  1052937.00  1021848.44  0.185242E+11  0.00000
 6.98    818.047   859.900  1022557.56  1000064.37  0.136823E+11  0.00000
 7.97    793.588   838.965   991984.37   975715.87  0.898264E+10  0.00000
 8.97    769.753   816.823   962190.69   949965.87  0.442377E+10  0.00000
 9.97    747.256   794.225   934068.44   923683.25  0.000000E+00  0.00000
QK 0.528555499E+11
QP,QG,(QG/TAUL) -598.291503      -12132789.0      -851.603271


      53.00 TAUH IN H.MIN

    X       TK        TG         UK          UG          ESP         SA
 0.00    992.859  1000.000  1520072.00  1163000.00  0.548359E+11  1.00000
 1.00    964.949   991.838  1485185.00  1153508.00  0.478246E+11  1.00000
 1.99    880.000   965.005  1369435.00  1122301.00  0.411649E+11  0.96572
 2.99    880.000   932.122  1211027.00  1084058.00  0.351447E+11  0.39795
 3.99    880.000   911.958  1141895.00  1060608.00  0.296555E+11  0.15017
 4.98    880.000   899.594  1105840.00  1046228.44  0.244118E+11  0.02094
 5.98    867.495   889.596  1084367.00  1034600.50  0.193023E+11  0.00000
 6.98    848.537   877.377  1060670.00  1020389.37  0.142982E+11  0.00000
 7.97    827.995   862.249  1034993.06  1002795.94  0.940920E+10  0.00000
 8.97    806.649   844.871  1008310.62   982585.00  0.464240E+10  0.00000
 9.97    785.334   825.950   981666.37   960580.56  0.000000E+00  0.00000
QK 0.548358635E+11
QP,QG,(QG/TAUL) -506.048828      -14101176.0      -792.433349
```

```
      54.00 TAUH IN H.MIN

    X        TK        TG          UK            UG           ESP          SA
 0.00    996.903  1000.000   1525127.00    1163000.00   0.565256E+11  1.00000
 1.00    981.744   995.869   1506179.00    1158196.00   0.494536E+11  1.00000
 1.99    931.122   980.613   1442901.00    1140453.00   0.425734E+11  1.00000
 2.99    880.000   951.578   1277779.00    1106685.00   0.362262E+11  0.63721
 3.99    880.000   923.890   1182713.00    1074484.00   0.304860E+11  0.29647
 4.98    880.000   906.908   1130834.00    1054734.00   0.250887E+11  0.11052
 5.98    880.000   896.506   1103667.00    1042636.69   0.198758E+11  0.01315
 6.98    869.058   837.992   1086321.00    1032734.62   0.147668E+11  0.00000
 7.97    853.203   877.616   1066503.00    1020667.87   0.974449E+10  0.00000
 8.97    835.643   864.748   1044552.06    1005702.06   0.481963E+10  0.00000
 9.97    817.116   849.961   1021394.31     988505.25   0.000000E+00  0.00000
QK  0.565255782E+11
QP,QG,(QG/TAUL) -436.236816      -15779101.0      -739.330810


      55.00 TAUH IN H.MIN

    X        TK        TG          UK            UG           ESP          SA
 0.00    998.648  1000.000   1527309.00    1163000.00   0.580051E+11  1.00000
 1.00    990.533   997.907   1517165.00    1160566.00   0.509024E+11  1.00000
 1.99    962.814   989.692   1482516.00    1151012.00   0.439042E+11  1.00000
 2.99    880.000   963.276   1362238.00    1120290.00   0.372674E+11  0.93993
 3.99    880.000   931.060   1234195.00    1082823.00   0.312100E+11  0.48099
 4.98    880.000   911.311   1162401.00    1059855.00   0.256189E+11  0.22367
 5.98    880.000   899.195   1122991.00    1045763.37   0.202873E+11  0.08241
 6.98    880.000   891.779   1101838.00    1037138.81   0.150970E+11  0.00659
 7.97    870.140   885.292   1087674.00    1029595.25   0.998913E+10  0.00000
 8.97    856.769   876.878   1070960.00    1019809.19   0.495327E+10  0.00000
 9.97    841.819   866.178   1052272.00    1007365.12   0.000000E+00  0.00000
QK  0.580051066E+11
QP,QG,(QG/TAUL) -389.087158      -17149168.0      -688.992919


      56.00 TAUH IN H.MIN

    X        TK        TG          UK            UG           ESP          SA
 0.00    999.406  1000.000   1528256.00    1163000.00   0.593461E+11  1.00000
 1.00    995.106   998.938   1522881.00    1161675.00   0.522279E+11  1.00000
 1.99    979.865   994.507   1503830.00    1156612.00   0.451666E+11  1.00000
 2.99    927.083   978.631   1437852.00    1138148.00   0.383037E+11  1.00000
 3.99    880.000   949.584   1298540.00    1104366.00   0.319198E+11  0.71162
 4.98    880.000   922.667   1201762.00    1073062.00   0.260867E+11  0.36475
 5.98    880.000   906.155   1147125.00    1053858.00   0.206069E+11  0.16891
 6.98    880.000   896.054   1116604.00    1042111.12   0.153259E+11  0.05952
 7.97    880.000   889.817   1100486.00    1034857.12   0.101536E+11  0.00175
 8.97    870.930   884.295   1088661.00    1028435.31   0.504659E+10  0.00000
 9.97    859.659   876.923   1074573.00    1019861.25   0.000000E+00  0.00000
QK  0.593461452E+11
QP,QG,(QG/TAUL) -357.847168      -18266992.0      -642.344970
```

```
***************************************************
DISCHARGING   TAU=   201600.
***************************************************


     57.00 TAUH IN H.MIN

     X        TK         TG        UK         UG         ESP          SA
   0.00    971.923    866.750  1493902.00   981160.94  0.572937E+11  1.00000
   1.00    963.169    843.364  1482960.00   954688.12  0.525728E+11  1.00000
   1.99    945.264    817.645  1460578.00   925574.62  0.477498E+11  1.00000
   2.99    899.139    793.388  1402923.00   898115.37  0.428195E+11  1.00000
   3.99    858.534    773.562  1271724.00   875671.69  0.377396E+11  0.71168
   4.98    853.939    754.378  1169197.00   853955.87  0.324272E+11  0.36478
   5.98    848.670    731.890  1107969.00   828499.62  0.267326E+11  0.16894
   6.98    842.335    705.535  1069528.00   798665.62  0.204927E+11  0.05953
   7.97    834.691    674.691  1043852.25   763750.87  0.138122E+11  0.00176
   8.97    818.829    639.425  1023535.06   723829.44  0.694497E+10  0.00000
   9.97    800.067    600.000  1000082.69   679200.00  0.000000E+00  0.00000
QK-0.593464238E+11
QP,QG,(QG/TAUL)  543.528564        2059223.00        572.197753


     58.00 TAUH IN H.MIN

     X        TK         TG        UK         UG         ESP          SA
   0.00    944.494    840.509  1459616.00   951456.69  0.554343E+11  1.00000
   1.00    932.459    817.780  1444573.00   925726.69  0.509707E+11  1.00000
   1.99    912.840    793.450  1420048.00   898185.19  0.463838E+11  1.00000
   2.99    872.052    770.483  1369063.00   872187.00  0.416646E+11  1.00000
   3.99    836.399    751.053  1244055.00   850192.75  0.367690E+11  0.71168
   4.98    828.312    732.190  1137163.00   828839.06  0.316172E+11  0.36478
   5.98    819.222    710.945  1071159.00   804789.44  0.260619E+11  0.16894
   6.98    808.575    687.063  1027328.19   777755.12  0.199655E+11  0.05953
   7.97    796.063    660.330   995567.94   747494.06  0.134586E+11  0.00176
   8.97    776.508    631.120   970633.62   714428.12  0.677544E+10  0.00000
   9.97    754.147    600.000   942682.50   679200.00  0.000000E+00  0.00000
QK-0.593464238E+11
QP,QG,(QG/TAUL)  490.061035        3918677.00        544.343261


     59.00 TAUH IN H.MIN

     X        TK         TG        UK         UG         ESP          SA
   0.00    917.560    816.159  1425949.00   923892.69  0.537618E+11  1.00000
   1.00    903.199    794.301  1407997.00   899148.31  0.495011E+11  1.00000
   1.99    882.610    771.427  1382261.00   873254.87  0.451095E+11  1.00000
   2.99    846.201    749.857  1336750.00   848838.25  0.405727E+11  1.00000
   3.99    814.459    731.189  1216631.00   827705.94  0.358431E+11  0.71168
   4.98    803.850    713.103  1106585.00   807232.19  0.308416E+11  0.36478
   5.98    792.140    693.412  1037306.56   784942.12  0.254216E+11  0.16894
   6.98    780.000    671.924   990034.69   760618.56  0.194646E+11  0.05389
   7.97    763.729    648.747   954660.75   734381.69  0.131212E+11  0.00000
   8.97    742.255    624.572   927817.37   707015.31  0.661157E+10  0.00000
   9.97    718.824    600.000   898528.75   679200.00  0.000000E+00  0.00000
QK-0.593464238E+11
QP,QG,(QG/TAUL)  440.445800        5591098.00        517.871826
```

```
      60.00 TAUH IN H.MIN
   X       TK       TG       UK          UG        ESP          SA
0.00    891.591  794.980  1393487.00   899917.44  0.522560E+11  1.00000
1.00    875.707  774.410  1373632.00   876632.00  0.481555E+11  1.00000
1.99    854.767  753.340  1347458.00   852780.69  0.439251E+11  1.00000
2.99    822.082  733.602  1306601.00   830437.87  0.395520E+11  1.00000
3.99    793.589  716.382  1190543.00   810944.69  0.349784E+11  0.71168
4.98    781.367  699.890  1078482.00   792275.69  0.301152E+11  0.36478
5.98    780.000  681.153  1006135.06   771065.37  0.248217E+11  0.11160
6.98    763.473  660.144   954340.37   747283.50  0.189959E+11  0.00000
7.97    736.150  639.348   920186.50   723741.87  0.128040E+11  0.00000
8.97    714.547  619.404   893182.44   701165.37  0.645566E+10  0.00000
9.97    691.623  600.000   864527.81   679200.00  0.000000E+00  0.00000
QK-0.593464238E+11
QP,QG,(QG/TAUL)   397.290527          7096589.00          493.034912


      61.00 TAUH IN H.MIN
   X       TK       TG       UK          UG        ESP          SA
0.00    867.145  777.798  1362930.00   880467.25  0.508910E+11  1.00000
1.00    850.449  759.009  1342060.00   859198.25  0.469169E+11  1.00000
1.99    829.717  740.202  1316145.00   837909.37  0.428194E+11  1.00000
2.99    800.364  722.844  1279453.00   818259.19  0.385871E+11  1.00000
3.99    780.000  707.227  1166785.00   800580.81  0.341668E+11  0.68741
4.98    780.000  690.350  1050074.00   781476.31  0.294483E+11  0.27157
5.98    777.441  669.854   971300.50   758274.44  0.242749E+11  0.00000
6.98    738.395  649.428   922992.81   735152.94  0.185678E+11  0.00000
7.97    712.964  631.744   891203.37   715134.12  0.125123E+11  0.00000
8.97    692.146  615.322   865181.12   696545.06  0.631073E+10  0.00000
9.97    670.658  600.000   838321.50   679200.00  0.000000E+00  0.00000
QK-0.593464238E+11
QP,QG,(QG/TAUL)   362.280273          8461239.00          470.235839


      62.00 TAUH IN H.MIN
   X       TK       TG       UK          UG        ESP          SA
0.00    844.636  762.413  1334794.00   863051.44  0.496451E+11  1.00000
1.00    827.728  745.301  1313659.00   843680.69  0.457707E+11  1.00000
1.99    807.663  728.508  1288578.00   824671.37  0.417826E+11  1.00000
2.99    781.443  713.189  1255802.00   807330.19  0.376695E+11  1.00000
3.99    780.000  697.529  1141476.00   789602.50  0.333884E+11  0.59669
4.98    780.000  678.405  1019377.19   767954.44  0.288198E+11  0.15906
5.98    751.167  658.186   938957.00   745067.12  0.237737E+11  0.00000
6.98    716.911  640.594   896138.06   725152.19  0.181859E+11  0.00000
7.97    693.551  625.603   866937.31   708182.56  0.122493E+11  0.00000
8.97    674.081  612.103   842599.94   692900.81  0.617883E+10  0.00000
9.97    654.517  600.000   818145.62   679200.00  0.000000E+00  0.00000
QK-0.593464238E+11
QP,QG,(QG/TAUL)   330.931884          9706767.00          449.574707
```

```
      63.00  TAUH IN H.MIN

     X      TK       TG        UK           UG          ESP           SA
  0.00   824.177  750.680  1309220.00   849770.12   0.484981E+11  1.00000
  1.00   807.533  735.563  1288415.00   832657.44   0.447026E+11  1.00000
  1.99   788.524  721.075  1264654.00   816257.19   0.408042E+11  1.00000
  2.99   780.000  706.423  1232925.00   799671.25   0.367922E+11  0.92447
  3.99   780.000  689.363  1113159.00   730359.50   0.326305E+11  0.49520
  4.98   780.000  668.348   984201.06   756570.00   0.282098E+11  0.03298
  5.98   728.616  648.413   910768.69   734004.00   0.233168E+11  0.00000
  6.98   698.503  633.304   873127.87   716900.12   0.178434E+11  0.00000
  7.97   677.322  620.639   846651.25   702563.44   0.120166E+11  0.00000
  8.97   659.507  609.560   824382.50   690021.87   0.606028E+10  0.00000
  9.97   642.069  600.000   802585.81   679200.00   0.000000E+00  0.00000
QK-0.593464238E+11
QP,QG,(QG/TAUL)   307.025390        10853252.0         430.839843


      64.00  TAUH IN H.MIN

     X      TK       TG        UK           UG          ESP           SA
  0.00   805.896  740.220  1286368.00   837929.25   0.474332E+11  1.00000
  1.00   789.893  726.843  1266365.00   822786.19   0.436998E+11  1.00000
  1.99   780.000  713.369  1243570.00   807533.44   0.398751E+11  0.96262
  2.99   780.000  697.918  1207616.00   790043.56   0.359482E+11  0.83375
  3.99   780.000  678.885  1081903.00   768498.00   0.318887E+11  0.38317
  4.98   758.734  657.901   948416.25   744743.62   0.276077E+11  0.00000
  5.98   709.296  640.248   886604.56   724760.56   0.228712E+11  0.00000
  6.98   682.813  627.304   853515.62   710108.37   0.175298E+11  0.00000
  7.97   663.810  616.634   829760.87   698030.06   0.118112E+11  0.00000
  8.97   647.777  607.553   809719.87   687750.50   0.595553E+10  0.00000
  9.97   632.480  600.000   790599.37   679200.00   0.000000E+00  0.00000
QK-0.593464238E+11
QP,QG,(QG/TAUL)   285.711669        11917694.0         413.976806


      65.00  TAUH IN H.MIN

     X      TK       TG        UK           UG          ESP           SA
  0.00   789.728  732.694  1266159.00   829409.94   0.464356E+11  1.00000
  1.00   780.000  720.597  1246986.00   815716.31   0.427513E+11  0.97487
  1.99   780.000  706.824  1220739.00   800124.69   0.389869E+11  0.88079
  2.99   780.000  689.857  1179464.00   780218.37   0.351317E+11  0.73285
  3.99   780.000  668.956  1046978.81   757258.62   0.311595E+11  0.25799
  4.98   734.338  648.499   917921.06   734101.19   0.269981E+11  0.00000
  5.98   692.684  633.421   865854.50   717033.00   0.224142E+11  0.00000
  6.98   669.476  622.368   836844.00   704250.56   0.172200E+11  0.00000
  7.97   652.577  613.402   815720.31   694371.37   0.116204E+11  0.00000
  8.97   638.336  605.969   797919.31   685956.87   0.586317E+10  0.00000
  9.97   625.085  600.000   781355.37   679200.00   0.000000E+00  0.00000
QK-0.593464238E+11
QP,QG,(QG/TAUL)   270.376953        12914757.0         398.779785
```

X	TK	TG	UK	UG	ESP	SA
0.00	780.000	728.143	1248794.00	824258.19	0.454800E+11	0.98135
1.00	780.000	716.121	1226850.00	810648.62	0.418345E+11	0.90268
1.99	780.000	701.310	1195904.00	793883.06	0.381194E+11	0.79178
2.99	780.000	683.066	1148871.00	773230.44	0.343243E+11	0.62320
3.99	780.000	660.591	1008884.69	747789.25	0.304267E+11	0.12146
4.98	713.709	640.587	892135.81	725145.12	0.263670E+11	0.00000
5.98	678.432	627.719	848038.75	710578.06	0.219321E+11	0.00000
6.98	658.166	618.308	822706.56	699924.44	0.168982E+11	0.00000
7.97	643.254	610.793	804066.44	691417.94	0.114279E+11	0.00000
8.97	630.740	604.717	788423.56	684539.31	0.577567E+10	0.00000
9.97	619.376	600.000	774219.81	679200.00	0.000000E+00	0.00000

QK-0.593464238E+11
QP,QG,(QG/TAUL)   261.103759   13869667.0   385.422607

X	TK	TG	UK	UG	ESP	SA
0.00	780.000	724.609	1231337.00	820258.06	0.445520E+11	0.91878
1.00	780.000	711.765	1205346.00	805718.50	0.409371E+11	0.82562
1.99	780.000	695.943	1169415.00	787808.12	0.372623E+11	0.69683
2.99	780.000	676.452	1116237.00	765744.19	0.335175E+11	0.50623
3.99	774.647	653.060	968307.94	739263.50	0.296832E+11	0.00000
4.98	696.297	633.947	870370.75	717627.75	0.257099E+11	0.00000
5.98	666.256	622.969	832818.94	705201.69	0.214204E+11	0.00000
6.98	648.615	614.976	810768.31	696153.37	0.165574E+11	0.00000
7.97	635.542	608.691	794426.25	689038.06	0.112250E+11	0.00000
8.97	624.640	603.728	780798.75	683420.50	0.568476E+10	0.00000
9.97	614.975	600.000	768717.81	679200.00	0.000000E+00	0.00000

QK-0.593464238E+11
QP,QG,(QG/TAUL)   253.903747   14796890.0   373.819335

X	TK	TG	UK	UG	ESP	SA
0.00	780.000	719.364	1212370.00	814319.69	0.436580E+11	0.85079
1.00	780.000	705.304	1181981.00	798404.56	0.400673E+11	0.74187
1.99	780.000	687.986	1140633.00	778799.94	0.364254E+11	0.59367
2.99	780.000	666.651	1080633.00	754648.62	0.327227E+11	0.37862
3.99	745.371	644.380	931712.44	729438.56	0.287481E+11	0.00000
4.98	681.494	628.366	851866.06	711311.00	0.250422E+11	0.00000
5.98	655.874	619.016	819841.19	700726.00	0.208813E+11	0.00000
6.98	640.565	612.243	800705.25	693059.06	0.161867E+11	0.00000
7.97	629.169	606.996	786460.12	687119.62	0.110046E+11	0.00000
8.97	619.741	602.948	774675.62	682536.62	0.558598E+10	0.00000
9.97	611.577	600.000	764470.62	679200.00	0.000000E+00	0.00000

QK-0.593464238E+11
QP,QG,(QG/TAUL)   243.214798   15690175.0   363.363525

```
      69.00 TAUH IN H.MIN

   X      TK        TG         UK           UG          ESP          SA
0.00   780.000   714.985  1191822.00   809363.56  0.427985E+11  0.77714
1.00   780.000   699.912  1156669.00   792301.00  0.392270E+11  0.65115
1.99   780.000   681.344  1109452.00   771281.62  0.356118E+11  0.48191
2.99   780.000   658.470  1041881.75   745388.81  0.319443E+11  0.23972
3.99   721.189   637.108   901485.87   721206.44  0.282092E+11  0.00000
4.98   668.867   623.676   836082.87   706000.75  0.243719E+11  0.00000
5.98   647.036   615.726   808794.25   697002.06  0.203184E+11  0.00000
6.98   633.790   610.000   792237.00   690520.75  0.157847E+11  0.00000
7.97   623.908   605.629   779883.94   685572.62  0.107657E+11  0.00000
8.97   615.807   602.330   769758.19   681837.81  0.547897E+10  0.00000
9.97   608.952   600.000   761189.00   679200.00  0.000000E+00  0.00000
QK-0.593464238E+11
QP,QG,(QG/TAUL)  234.293869       16548876.0         353.757324


      70.00 TAUH IN H.MIN

   X      TK        TG         UK           UG          ESP          SA
0.00   780.000   711.337  1169982.00   805233.62  0.419690E+11  0.69887
1.00   780.000   695.419  1129765.00   787214.12  0.384126E+11  0.55472
1.99   780.000   675.809  1076310.00   765016.19  0.348193E+11  0.36312
2.99   780.000   651.653  1000555.69   737671.81  0.311809E+11  0.09160
3.99   701.222   631.026   876526.19   714321.56  0.274843E+11  0.00000
4.98   658.132   619.743   822663.37   701549.62  0.237001E+11  0.00000
5.98   639.541   612.996   799424.81   693911.69  0.197361E+11  0.00000
6.98   628.110   608.165   785136.25   688443.25  0.153570E+11  0.00000
7.97   619.578   604.529   774471.19   684327.44  0.105119E+11  0.00000
8.97   612.654   601.843   765817.00   681286.00  0.536523E+10  0.00000
9.97   606.926   600.000   758656.12   679200.00  0.000000E+00  0.00000
QK-0.593464238E+11
QP,QG,(QG/TAUL)  226.859802       17371216.0         344.819580


      71.00 TAUH IN H.MIN

   X      TK        TG         UK           UG          ESP          SA
0.00   780.000   706.768  1146950.00   800061.37  0.411663E+11  0.61631
1.00   780.000   689.788  1101393.00   780839.87  0.376218E+11  0.45303
1.99   780.000   668.871  1041340.37   757161.94  0.340462E+11  0.23778
2.99   766.347   644.683   957933.12   729781.00  0.304321E+11  0.00000
3.99   684.688   625.937   855858.62   703560.87  0.267679E+11  0.00000
4.98   649.012   616.449   811263.69   697820.00  0.230289E+11  0.00000
5.98   633.191   610.731   791488.37   691347.87  0.191396E+11  0.00000
6.98   623.352   606.663   779189.31   686742.69  0.149083E+11  0.00000
7.97   616.016   603.644   770018.87   683324.69  0.102441E+11  0.00000
8.97   610.126   601.457   762657.25   680849.87  0.524528E+10  0.00000
9.97   605.360   600.000   756698.94   679200.00  0.000000E+00  0.00000
QK-0.593464238E+11
QP,QG,(QG/TAUL)  217.549865       18160336.0         336.457763


      72.00 TAUH IN H.MIN

   X      TK        TG         UK           UG          ESP          SA
0.00   780.000   700.943  1122059.00   793468.00  0.404054E+11  0.52710
1.00   780.000   682.615  1070731.00   772720.12  0.368703E+11  0.34313
1.99   780.000   660.035  1003361.56   747159.87  0.333092E+11  0.10166
2.99   737.224   637.176   921528.87   721283.81  0.297152E+11  0.00000
3.99   670.864   621.669   838578.44   703730.06  0.260782E+11  0.00000
4.98   641.276   613.691   801594.44   694698.81  0.223774E+11  0.00000
5.98   627.824   608.854   784778.62   639222.81  0.185511E+11  0.00000
6.98   619.375   605.434   774217.87   685351.81  0.144449E+11  0.00000
7.97   613.090   602.930   766361.31   682517.25  0.995435E+10  0.00000
8.97   608.101   601.153   760125.37   680505.12  0.511566E+10  0.00000
9.97   604.150   600.000   755186.19   679200.00  0.000000E+00  0.00000
QK-0.593464238E+11
QP,QG,(QG/TAUL)  205.681930       18909888.0         328.439209
```

NUMERICAL OUTPUT No. II

```
        11         =M NUMBER OF CALCULATION POINTS IN X-DIRECTION
1000.00000         =TGEBE [C] INLET TEMPERATURE FLUID (CHARGING)
 600.000000        =TGEEN [C] INLET TEMPERATURE FLUID (DISCHARGING)
 880.000000        =TKS [C] MELTING POINT
 780.000000        =TKK [C] CRYSTALLIZATION POINT
  20.0000000       =TU [C] AMBIENT TEMPERATUR
2500.00000         =RHOK [KG/M**3] DENSITY SOLID MEDIUM
1100.00000         =CKK [J/(KG*K)] SPECIFIC HEAT CERAMIC
1433.33007         =CKS [J/(KG*K)] SPECIFIC HEAT PHASE-CHANGE-MATERIAL
   5.00000000      =RLAMK [W/(M*K)] COEFF. OF THERMAL CONDUCT. SOLID MED.
 620000.000        =RK [J/KG] HEAT OF FUSION
0.449999988        =RMS MASS-SHARE OF SALT
0.310500025        =RHOGB [KG/M**3] DENSITY FLUID (CHARGING)
   3.66800022      =RHOGE [KG/M**3] DENSITY FLUID (DISCHARGING)
1163.00000         =CGB [J/(KG*K)] SPECIFIC HEAT FLUID (CHARGING)
1132.00000         =CGE [J/(KG*K)] SPECIFIC HEAT FLUID (DISCHARGING)
0.727999806E-01    =RLAMGB [W/(M*K)] COEFF. OF THERMAL COND. FLUID CHARG.
0.665000081E-01    =RLAMGE [W/(M*K)] COEFF. OF THERM. COND. FLUID DISCH.
0.709999978        =PRGB PRANDTL-NUMBER (CHARGING)
  69.0000000       =PRGE PRANDTL-NUMBER (DISCHARGING)
 800.000000        =TKVA NOMINAL TEMPERATURE FOR HEAT LOSS
   1.00000000      =DTKVU TEMPERATURE GRADIENT FOR HEAT LOSS
 300000.000        =QPDIS[W] ENERGY OUTPUT (DISCHARGING)
  57600.0000       =TAUDIS [S] DISCHARGING-TIME
  28800.0000       =TAUCHA [S] CHARGING-TIME
0.699999988        =ETASTO EFFICIENCY OF STORING (STARTVALUE)
 700.000000        =TCOUT [C] OUTLET-TEMPERATURE FLUID (CHARGING)
 750.000000        =TDOUT [C] OUTLET-TEMPERATURE FLUID (DISCHARGING)
0.699999999        =FF FILLING FACTOR
0.143700002E-03    =RNUEB [M**2/S] VISCOSITY FLUID (CHARGING)
0.110500003E-02    =RNUEE [M**2/S] VISCOSITY FLUID (DISCHARGING)
 200.000000        =DELTAT [C] AVERAGE TEMPERATURE DIFFERENCE CERAMIC


AB= 0.416178107
AE= 0.114388525
B= 0.199999986E-02
C= 0.118000009E-07
DB= 7808.19140
DE= 181.671188
EB= 0.234460413
EE= 0.181297697E-01
EST 0.246857113E+11
EMST 46664.8984
VS 18.6659545
QPCHA -0.230789184E-03
RMPGB 2.50000000
RMPGE 1.79999923
H 9.46819782
DA 1.89363956
DKRIT 0.349841937E-01
DI 0.349999964E-01
RET 2321.04687
REL 18.3968200
RNUT 10.5602321
RNUL 3.09793567
ALPHAT 21.9652710
ALPHAL 5.88607788
ALPHAB 21.2138824
ALPHAE 5.83073806
NBOHR 879
AG 0.845696687
AK 1.97064113
U 96.6510467
S 0.483749248E-01
```

```
TAU=    3601.7   TOUT=   694.00   SUMMET= 0.239725E+07
TAU=    7203.3   TOUT=   742.41   SUMMET= 0.496739E+07
TAU=   10805.8   TOUT=   786.05   SUMMET= 0.770367E+07
TAU=   14407.7   TOUT=   822.15   SUMMET= 0.105846E+08
TAU=   18009.3   TOUT=   849.33   SUMMET= 0.135802E+08
TAU=   21611.0   TOUT=   867.35   SUMMET= 0.166556E+08
TAU=   25212.2   TOUT=   877.63   SUMMET= 0.195613E+08
TAU=   28514.2   TOUT=   882.15   SUMMET= 0.222172E+08
****************************************************************
DISCHARGING  TAU=   28800.
****************************************************************
TAU=   32401.2   TOUT=   910.76   SUMMET= 0.333726E+07
TAU=   36004.0   TOUT=   880.08   SUMMET= 0.656172E+07
TAU=   39606.7   TOUT=   851.76   SUMMET= 0.967941E+07
TAU=   43209.5   TOUT=   826.97   SUMMET= 0.127010E+08
TAU=   46812.3   TOUT=   804.85   SUMMET= 0.156384E+08
TAU=   50415.1   TOUT=   786.45   SUMMET= 0.184925E+08
TAU=   54017.9   TOUT=   771.33   SUMMET= 0.212784E+08
TAU=   57620.7   TOUT=   758.91   SUMMET= 0.240164E+08
TAU=   61223.4   TOUT=   750.87   SUMMET= 0.267172E+08
TAU=   64826.2   TOUT=   746.62   SUMMET= 0.294021E+08
TAU=   68429.0   TOUT=   744.32   SUMMET= 0.320701E+08
TAU=   72031.8   TOUT=   740.11   SUMMET= 0.347197E+08
TAU=   75634.6   TOUT=   736.81   SUMMET= 0.373693E+08
TAU=   79237.4   TOUT=   731.04   SUMMET= 0.399927E+08
TAU=   82840.1   TOUT=   726.23   SUMMET= 0.426070E+08
TAU=   86142.7   TOUT=   718.98   SUMMET= 0.449746E+08
****************************************************************
CHARGING    TAU=   86400.
****************************************************************
TAU=   90002.1   TOUT=   643.49   SUMMET= 0.224227E+07
TAU=   93604.3   TOUT=   689.84   SUMMET= 0.462362E+07
TAU=   97206.0   TOUT=   739.58   SUMMET= 0.717994E+07
TAU=  100807.7   TOUT=   785.04   SUMMET= 0.990898E+07
TAU=  104409.3   TOUT=   821.52   SUMMET= 0.127877E+08
TAU=  108011.0   TOUT=   847.90   SUMMET= 0.157793E+08
TAU=  111612.7   TOUT=   865.16   SUMMET= 0.187241E+08
TAU=  114914.2   TOUT=   874.58   SUMMET= 0.213799E+08
****************************************************************
DISCHARGING  TAU=   115200.
****************************************************************
TAU=  118801.2   TOUT=   896.33   SUMMET= 0.328454E+07
TAU=  122404.0   TOUT=   866.27   SUMMET= 0.645811E+07
TAU=  126006.7   TOUT=   838.35   SUMMET= 0.952661E+07
TAU=  129609.5   TOUT=   813.54   SUMMET= 0.124998E+08
TAU=  133212.0   TOUT=   792.91   SUMMET= 0.153918E+08
TAU=  136815.1   TOUT=   775.00   SUMMET= 0.182062E+08
TAU=  140417.9   TOUT=   761.45   SUMMET= 0.209527E+08
TAU=  144020.7   TOUT=   750.19   SUMMET= 0.236565E+08
TAU=  147623.4   TOUT=   742.79   SUMMET= 0.263261E+08
TAU=  151226.2   TOUT=   738.63   SUMMET= 0.289757E+08
TAU=  154829.0   TOUT=   735.93   SUMMET= 0.316253E+08
TAU=  158431.8   TOUT=   732.00   SUMMET= 0.342471E+08
TAU=  162034.6   TOUT=   727.04   SUMMET= 0.368613E+08
TAU=  165637.3   TOUT=   721.61   SUMMET= 0.394521E+08
TAU=  169240.1   TOUT=   714.40   SUMMET= 0.420234E+08
TAU=  172542.7   TOUT=   707.65   SUMMET= 0.443551E+08
****************************************************************
CHARGING    TAU=   172800.
****************************************************************
TAU=  176402.1   TOUT=   636.61   SUMMET= 0.222620E+07
TAU=  180003.7   TOUT=   679.86   SUMMET= 0.457568E+07
TAU=  183605.4   TOUT=   728.81   SUMMET= 0.709395E+07
TAU=  187207.1   TOUT=   775.20   SUMMET= 0.978535E+07
TAU=  190808.7   TOUT=   813.45   SUMMET= 0.126310E+08
TAU=  194410.4   TOUT=   842.01   SUMMET= 0.155976E+08
TAU=  198012.1   TOUT=   861.14   SUMMET= 0.185471E+08
TAU=  201313.6   TOUT=   872.10   SUMMET= 0.212029E+08
****************************************************************
DISCHARGING  TAU=   201600.
****************************************************************
TAU=  205201.2   TOUT=   893.42   SUMMET= 0.327363E+07
TAU=  208804.0   TOUT=   863.60   SUMMET= 0.643713E+07
TAU=  212406.7   TOUT=   835.84   SUMMET= 0.949639E+07
TAU=  216009.5   TOUT=   811.23   SUMMET= 0.124607E+08
TAU=  219612.3   TOUT=   790.53   SUMMET= 0.153440E+08
TAU=  223215.1   TOUT=   772.91   SUMMET= 0.181510E+08
TAU=  226817.9   TOUT=   758.98   SUMMET= 0.208925E+08
TAU=  230420.7   TOUT=   747.92   SUMMET= 0.235881E+08
TAU=  234023.4   TOUT=   741.37   SUMMET= 0.262479E+08
TAU=  237626.2   TOUT=   737.72   SUMMET= 0.288975E+08
TAU=  241229.0   TOUT=   733.56   SUMMET= 0.315359E+08
TAU=  244831.8   TOUT=   728.44   SUMMET= 0.341502E+08
TAU=  248434.6   TOUT=   724.35   SUMMET= 0.367525E+08
TAU=  252037.3   TOUT=   718.03   SUMMET= 0.393315E+08
TAU=  255640.1   TOUT=   711.57   SUMMET= 0.418863E+08
TAU=  258942.7   TOUT=   703.44   SUMMET= 0.442092E+08
```

```
       48.00 TAUH IN H.MIN
     X        TK        TG         UK           UG         ESP         SA
    0.00    780.000   706.868   1112340.00    800174.44  0.396011E+11  0.49226
    0.95    780.000   684.828   1047568.87    775225.56  0.360874E+11  0.26011
    1.89    768.105   657.934    960129.94    744781.12  0.325572E+11  0.00000
    2.84    706.971   633.940    883712.69    717620.31  0.290044E+11  0.00000
    3.79    663.118   618.536    828896.56    700182.87  0.254198E+11  0.00000
    4.73    629.440   610.174    786798.94    690717.31  0.217832E+11  0.00000
    5.68    617.952   606.098    772439.25    686103.62  0.180150E+11  0.00000
    6.63    611.569   603.488    764460.69    683148.19  0.140206E+11  0.00000
    7.57    607.091   601.727    758862.37    681154.69  0.972024E+10  0.00000
    8.52    603.785   600.608    754730.69    679888.62  0.503765E+10  0.00000
    9.47    601.469   600.000    751834.94    679200.00  0.000000E+00  0.00000
 QK-0.603030487E+11
 QP,QG,(QG/TAUL)  217.753311        20617280.0        358.506591


 *******************************************************
 CHARGING     TAU=  172800.
 *******************************************************

       49.00 TAUH IN H.MIN
     X        TK         TG         UK           UG          ESP         SA
    0.00    880.000   1000.000   1325993.00   1163000.00  0.435671E+11  0.81002
    0.95    880.000    939.929   1176874.00   1093138.00  0.377295E+11  0.27554
    1.89    845.896    901.394   1057368.00   1048321.56  0.325185E+11  0.00000
    2.84    795.661    861.040    994575.31   1001389.12  0.277328E+11  0.00000
    3.79    748.880    816.601    936099.25    949706.62  0.232299E+11  0.00000
    4.73    707.002    772.222    883751.06    898094.44  0.189854E+11  0.00000
    5.68    676.581    731.951    845725.50    851258.81  0.149518E+11  0.00000
    6.63    654.148    698.635    817684.56    812513.19  0.110722E+11  0.00000
    7.57    637.674    672.210    797090.94    781780.94  0.730607E+10  0.00000
    8.52    625.669    651.963    782084.94    758233.25  0.362296E+10  0.00000
    9.47    617.040    636.590    771299.44    740353.87  0.000000E+00  0.00000
 QK 0.435670712E+11
 QP,QG,(QG/TAUL) -1056.61499       -3952822.00       -1099.31030


       50.00 TAUH IN H.MIN
     X        TK         TG         UK           UG          ESP         SA
    0.00    948.332   1000.000   1464414.00   1163000.00  0.471544E+11  1.00000
    0.95    880.000    957.032   1277097.00   1113028.00  0.407601E+11  0.63476
    1.89    880.000    918.471   1118783.00   1068182.00  0.351721E+11  0.06733
    2.84    856.800    893.407   1070998.00   1039032.37  0.300649E+11  0.00000
    3.79    818.379    865.465   1022972.06   1006535.94  0.251811E+11  0.00000
    4.73    778.989    832.036    973734.56    967657.75  0.205242E+11  0.00000
    5.68    743.194    796.517    928990.87    926349.94  0.160865E+11  0.00000
    6.63    712.135    762.052    890167.31    886266.37  0.118436E+11  0.00000
    7.57    686.129    730.561    857660.44    849642.12  0.776719E+10  0.00000
    8.52    664.959    703.009    831197.94    817599.94  0.382827E+10  0.00000
    9.47    648.172    679.765    810214.44    790567.31  0.000000E+00  0.00000
 QK 0.471543726E+11
 QP,QG,(QG/TAUL) -931.082031       -7521486.00       -1053.51928
```

```
   51.00 TAUH IN H.MIN

   X        TK       TG         UK         UG         ESP          SA
0.00    984.077 1000.000  1509095.00 1163000.00 0.502525E+11  1.00000
0.95    896.471  970.101  1399587.00 1128227.00 0.434683E+11  1.00000
1.89    880.000  929.119  1181195.00 1080565.00 0.374489E+11  0.29103
2.84    880.000  904.531  1110175.00 1051969.00 0.321047E+11  0.03648
3.79    861.583  887.641  1076978.00 1032327.00 0.270036E+11  0.00000
4.73    831.913  867.170  1039889.37 1008518.81 0.220665E+11  0.00000
5.68    800.396  841.634  1000494.31  978820.62 0.173077E+11  0.00000
6.63    769.230  813.183   961536.25  945731.87 0.127316E+11  0.00000
7.57    739.976  783.869   924969.06  911640.37 0.833173E+10  0.00000
8.52    713.604  755.291   892003.37  878403.81 0.409401E+10  0.00000
9.47    690.680  728.669   863349.31  847441.94 0.000000E+00  0.00000
QK 0.502524682E+11
QP,QG,(QG/TAUL) -788.895507      -10600136.0      -993.038085


   52.00 TAUH IN H.MIN

   X        TK       TG         UK         UG         ESP          SA
0.00    994.990 1000.000  1522736.00 1163000.00 0.528450E+11  1.00000
0.95    956.144  987.769  1474178.00 1148775.00 0.458550E+11  1.00000
1.89    880.000  952.879  1275551.00 1108198.00 0.394416E+11  0.62922
2.84    880.000  916.397  1157045.00 1065770.00 0.337679E+11  0.20447
3.79    880.000  898.179  1107230.00 1044581.94 0.284870E+11  0.02592
4.73    865.334  885.408  1081666.00 1029729.94 0.233818E+11  0.00000
5.68    841.900  869.491  1052374.00 1011218.37 0.184046E+11  0.00000
6.63    815.901  849.169  1019874.75  987584.12 0.135715E+11  0.00000
7.57    788.894  825.773   986115.69  960373.69 0.889289E+10  0.00000
8.52    762.214  800.597   952766.00  931094.87 0.437083E+10  0.00000
9.47    737.123  775.123   921279.19  901467.87 0.000000E+00  0.00000
QK 0.528450150E+11
QP,QG,(QG/TAUL) -653.830322      -13174176.0      -927.137939


   53.00 TAUH IN H.MIN

   X        TK       TG         UK         UG         ESP          SA
0.00    998.397 1000.000  1526995.00 1163000.00 0.549906E+11  1.00000
0.95    981.320  994.923  1505648.00 1157095.00 0.479173E+11  1.00000
1.89    891.323  965.587  1393152.00 1122978.00 0.411561E+11  1.00000
2.84    880.000  935.574  1216203.00 1076443.00 0.350701E+11  0.41651
3.79    880.000  903.762  1136727.00 1049913.00 0.295823E+11  0.13165
4.73    880.000  891.369  1102476.00 1036662.06 0.243598E+11  0.00888
5.68    867.114  882.451  1083891.00 1026291.25 0.192605E+11  0.00000
6.63    848.555  870.127  1060693.00 1011957.37 0.142587E+11  0.00000
7.57    827.118  853.973  1033896.56  993170.31 0.937352E+10  0.00000
8.52    803.930  834.710  1004911.75  970767.62 0.461840E+10  0.00000
9.47    780.226  813.380   975281.75  945961.44 0.000000E+00  0.00000
QK 0.549905817E+11
QP,QG,(QG/TAUL) -542.596435      -15301539.0      -862.317627
```

```
      54.00 TAUH IN H.MIN

     X       TK        TG         UK          UG         ESP        SA
  0.00    999.477  1000.000  1528344.00  1163000.00  0.567859E+11  1.00000
  0.95    992.020   997.871  1519023.00  1160524.00  0.496783E+11  1.00000
  1.89    952.191   984.973  1469237.00  1145524.00  0.427085E+11  1.00000
  2.84    880.000   950.491  1306075.00  1105421.00  0.362354E+11  0.73863
  3.79    880.000   915.209  1181427.00  1064388.00  0.304336E+11  0.29186
  4.73    880.000   897.579  1124757.00  1043885.06  0.250549E+11  0.08874
  5.68    880.000   888.783  1100998.00  1033655.06  0.198637E+11  0.00358
  6.63    868.907   881:608  1086133.00  1025310.69  0.147627E+11  0.00000
  7.57    853.707   871.439  1067132.00  1013483.56  0.974063E+10  0.00000
  8.52    835.746   858.084  1044681.69   997952.00  0.481525E+10  0.00000
  9.47    815.929   841.919  1019910.44   979152.12  0.000000E+00  0.00000
QK 0.567859322E+11
QP,QG,(QG/TAUL) -459.619873        -16987904.0        -798.313964


      55.00 TAUH IN H.MIN

     X       TK        TG         UK          UG         ESP        SA
  0.00    999.824  1000.000  1528779.00  1163000.00  0.583335E+11  1.00000
  0.95    996.604   999.105  1524753.00  1161959.00  0.512115E+11  1.00000
  1.89    978.772   993.388  1502464.00  1155311.00  0.441508E+11  1.00000
  2.84    909.186   968.655  1415481.00  1126546.00  0.373450E+11  1.00000
  3.79    880.000   931.581  1241866.00  1083429.00  0.311471E+11  0.50849
  4.73    880.000   905.759  1154920.00  1053398.00  0.255570E+11  0.19685
  5.68    880.000   892.864  1116021.00  1038400.87  0.202604E+11  0.05743
  6.63    879.535   886.311  1099417.00  1030779.81  0.150933E+11  0.00000
  7.57    870.010   880.530  1087511.00  1024056.12  0.999277E+10  0.00000
  8.52    857.520   872.143  1071899.00  1014302.06  0.495638E+10  0.00000
  9.47    842.564   861.070  1053204.00  1001424.87  0.000000E+00  0.00000
QK 0.583335198E+11
QP,QG,(QG/TAUL) -403.937988        -18146272.0        -731.268554


      56.00 TAUH IN H.MIN

     X       TK        TG         UK          UG         ESP        SA
  0.00    999.940  1000.000  1528923.00  1163000.00  0.597206E+11  1.00000
  0.95    998.556   999.622  1527194.00  1162561.00  0.525926E+11  1.00000
  1.89    990.568   997.090  1517209.00  1159616.00  0.454918E+11  1.00000
  2.94    958.583   985.815  1477228.00  1146503.00  0.385076E+11  1.00000
  3.79    880.000   952.522  1337368.00  1107783.00  0.319429E+11  0.85079
  4.73    880.000   916.205  1202386.00  1065547.00  0.260193E+11  0.36698
  5.68    880.000   898.105  1139727.00  1044496.12  0.205567E+11  0.14240
  6.63    880.000   889.003  1111232.00  1033910.75  0.153067E+11  0.04027
  7.57    879.103   884.329  1098877.00  1028424.19  0.101521E+11  0.00000
  8.52    871.176   879.661  1088968.00  1023046.19  0.504939E+10  0.00000
  9.47    860.815   872.883  1076017.00  1015163.50  0.000000E+00  0.00000
QK 0.597206425E+11
QP,QG,(QG/TAUL) -369.591552        -19304928.0        -680.930175
```

```
***************************************************
DISCHARGING  TAU=  201600.
***************************************************


   57.00 TAUH IN H.MIN

   X        TK       TG        UK          UG          ESP          SA
 0.00    974.112   893.431  1496639.00  1011363.94  0.574579E+11  1.00000
 0.95    966.672   870.225  1487338.00   985095.19  0.528001E+11  1.00000
 1.89    953.140   843.187  1470424.00   954487.37  0.480245E+11  1.00000
 2.84    921.830   814.757  1431286.00   922304.62  0.431380E+11  1.00000
 3.79    859.504   791.870  1311749.00   896396.56  0.381029E+11  0.85079
 4.73    853.866   772.325  1169719.00   874272.00  0.328120E+11  0.36698
 5.68    847.274   748.733  1098820.00   847566.06  0.270243E+11  0.14240
 6.63    839.031   720.267  1060021.00   815342.81  0.206264E+11  0.04027
 7.57    828.105   686.112  1035130.12   776679.06  0.138585E+11  0.00000
 8.52    809.947   646.049  1012432.50   731327.50  0.695981E+10  0.00000
 9.47    787.723   600.000   984652.69   679200.00  0.000000E+00  0.00000
QK-0.597206425E+11
QP,QG,(QG/TAUL)   597.894043        2265694.00        629.480712


   58.00 TAUH IN H.MIN

   X        TK       TG        UK          UG          ESP          SA
 0.00    946.972   863.642  1462713.00   977642.50  0.554164E+11  1.00000
 0.95    935.098   840.307  1447871.00   951227.94  0.510512E+11  1.00000
 1.89    917.875   814.325  1426342.00   921815.69  0.465401E+11  1.00000
 2.84    887.582   787.672  1388476.00   891644.31  0.418866E+11  1.00000
 3.79    836.776   765.208  1283339.00   866215.94  0.370520E+11  0.85079
 4.73    826.862   745.124  1135964.00   843480.06  0.319322E+11  0.36698
 5.68    815.569   722.182  1059189.00   817510.50  0.262895E+11  0.14240
 6.63    801.983   696.077  1013711.31   787958.81  0.200577E+11  0.04027
 7.57    785.225   666.677   981530.62   754678.62  0.134924E+11  0.00000
 8.52    762.125   634.426   952654.37   718170.12  0.678864E+10  0.00000
 9.47    735.171   600.000   918763.06   679200.00  0.000000E+00  0.00000
QK-0.597206425E+11
QP,QG,(QG/TAUL)   537.195556        4306506.00        598.666748


   59.00 TAUH IN H.MIN

   X        TK       TG        UK          UG          ESP          SA
 0.00    919.317   835.885  1428145.00   946222.50  0.535861E+11  1.00000
 0.95    904.320   812.992  1409398.00   920306.75  0.494396E+11  1.00000
 1.89    884.793   788.401  1384990.00   892469.81  0.451409E+11  1.00000
 2.84    855.767   763.715  1348708.00   864525.31  0.406883E+11  1.00000
 3.79    813.336   742.358  1254039.00   840349.75  0.360367E+11  0.85079
 4.73    800.422   722.905  1102914.00   818328.62  0.310799E+11  0.36698
 5.68    786.078   701.697  1022325.44   794321.31  0.255826E+11  0.14240
 6.63    777.687   677.524   972106.87   766957.31  0.195120E+11  0.00000
 7.57    749.594   651.565   936991.00   737572.06  0.131359E+11  0.00000
 8.52    724.906   625.739   906131.19   708336.06  0.661828E+10  0.00000
 9.47    697.368   600.000   871709.00   679200.00  0.000000E+00  0.00000
QK-0.597206425E+11
QP,QG,(QG/TAUL)   480.639648        6136082.00        568.761962
```

```
     60.00 TAUH IN H.MIN

   X      TK        TG        UK          UG          ESP          SA
0.00   892.048   811.289   1394059.00   918379.25   0.519498E+11  1.00000
0.95   875.037   789.506   1372795.00   893721.25   0.479667E+11  1.00000
1.89   854.323   766.844   1346903.00   868068.00   0.438375E+11  1.00000
2.84   826.832   744.617   1312539.00   842906.31   0.395595E+11  1.00000
3.79   790.764   725.270   1225824.00   821006.00   0.350816E+11  0.85079
4.73   780.000   707.152   1072397.00   800496.37   0.302805E+11  0.34910
5.68   780.000   685.198    986099.44   775643.62   0.249202E+11  0.03979
6.63   747.089   661.582    933859.87   748910.62   0.189997E+11  0.00000
7.57   720.292   639.845    900364.56   724305.19   0.127969E+11  0.00000
8.52   696.056   619.250    870069.06   700991.19   0.645342E+10  0.00000
9.47   670.186   600.000    837731.12   679200.00   0.000000E+00  0.00000
QK-0.597206425E+11
QP,QG,(QG/TAUL)  430.521972      7771538.00       540.370849


     61.00 TAUH IN H.MIN

   X      TK        TG        UK          UG          ESP          SA
0.00   866.148   790.604   1361684.00   894963.62   0.504782E+11  1.00000
0.95   848.124   770.545   1339153.00   872257.06   0.466172E+11  1.00000
1.89   827.205   750.312   1313005.00   849453.56   0.426229E+11  1.00000
2.84   801.626   730.988   1281031.00   827478.69   0.384903E+11  1.00000
3.79   780.000   712.957   1199849.00   807067.12   0.341788E+11  0.80592
4.73   780.000   692.750   1039572.87   784193.69   0.295453E+11  0.23145
5.68   757.678   669.815    947095.94   758231.00   0.243221E+11  0.00000
6.63   721.203   648.827    901502.19   734472.62   0.185357E+11  0.00000
7.57   696.328   630.760    870409.19   714200.37   0.124854E+11  0.00000
8.52   673.743   614.401    842177.25   695501.56   0.629944E+10  0.00000
9.47   650.620   600.000    813274.25   679200.00   0.000000E+00  0.00000
QK-0.597206425E+11
QP,QG,(QG/TAUL)  388.373779      9242245.00       514.165771


     62.00 TAUH IN H.MIN

   X      TK        TG        UK          UG          ESP          SA
0.00   842.171   772.979   1331712.00   875012.81   0.491460E+11  1.00000
0.95   823.948   754.867   1308934.00   854509.37   0.453763E+11  1.00000
1.89   803.586   737.112   1283481.00   834410.62   0.414891E+11  1.00000
2.84   780.285   720.585   1254355.00   815702.50   0.374769E+11  1.00000
3.79   780.000   702.635   1170123.00   795383.06   0.333069E+11  0.69937
4.73   780.000   679.320   1000488.37   768990.37   0.288424E+11  0.09136
5.68   730.969   656.363    913710.06   743003.25   0.237798E+11  0.00000
6.63   699.375   638.636    874217.44   722935.75   0.181249E+11  0.00000
7.57   676.846   623.723    846057.12   706054.12   0.122056E+11  0.00000
8.52   656.519   610.774    820647.50   691396.19   0.615906E+10  0.00000
9.47   636.524   600.000    795654.69   679200.00   0.000000E+00  0.00000
QK-0.597206425E+11
QP,QG,(QG/TAUL)  352.462402      10573501.0       490.189697
```

63.00 TAUH IN H.MIN

X	TK	TG	UK	UG	ESP	SA
0.00	820.620	759.049	1304774.00	859243.50	0.479279E+11	1.00000
0.95	802.875	743.162	1282593.00	841259.25	0.442264E+11	1.00000
1.89	783.725	728.047	1258655.00	824149.87	0.404237E+11	1.00000
2.84	780.000	711.827	1228412.00	805788.31	0.365105E+11	0.90829
3.79	780.000	691.281	1136200.00	782530.25	0.324592E+11	0.57778
4.73	765.862	666.669	957326.00	754669.87	0.281609E+11	0.00000
5.68	708.513	645.414	885640.12	730608.25	0.232781E+11	0.00000
6.63	681.128	630.520	851408.56	713749.06	0.177629E+11	0.00000
7.57	661.111	618.282	826387.25	699896.00	0.119620E+11	0.00000
8.52	643.264	608.063	804078.75	688328.06	0.603479E+10	0.00000
9.47	626.373	600.000	782965.75	679200.00	0.000000E+00	0.00000

QK-0.597206425E+11
QP,QG,(QG/TAUL)   324.077392   11790704.0   468.562011

64.00 TAUH IN H.MIN

X	TK	TG	UK	UG	ESP	SA
0.00	801.498	747.973	1280871.00	846705.62	0.468050E+11	1.00000
0.95	784.721	734.368	1259900.00	831304.50	0.431546E+11	1.00000
1.89	780.000	719.903	1235409.00	814930.62	0.394185E+11	0.93337
2.84	780.000	701.792	1198212.00	794429.12	0.355105E+11	0.80005
3.79	780.000	678.223	1096657.00	767748.87	0.316345E+11	0.43605
4.73	736.000	654.178	919999.12	740529.44	0.274784E+11	0.00000
5.68	689.584	636.510	861979.69	720529.06	0.227749E+11	0.00000
6.63	665.971	624.070	832462.94	706447.19	0.174224E+11	0.00000
7.57	648.455	614.080	810567.37	695138.44	0.117462E+11	0.00000
8.52	633.077	606.036	791345.12	686033.37	0.592612E+10	0.00000
9.47	619.055	600.000	773817.81	679200.00	0.000000E+00	0.00000

QK-0.597206425E+11
QP,QG,(QG/TAUL)   301.509521   12912617.0   449.025878

65.00 TAUH IN H.MIN

X	TK	TG	UK	UG	ESP	SA
0.00	785.310	741.402	1260636.00	839267.00	0.457466E+11	1.00000
0.95	780.000	728.969	1239688.00	825193.50	0.421344E+11	0.94871
1.89	780.000	713.591	1209316.00	807785.37	0.384506E+11	0.83985
2.84	780.000	693.579	1164249.00	785131.56	0.346846E+11	0.67832
3.79	780.000	667.536	1052004.00	755650.81	0.308155E+11	0.27600
4.73	711.550	643.955	889436.56	728956.75	0.267771E+11	0.00000
5.68	673.687	629.286	842107.37	712352.06	0.222490E+11	0.00000
6.63	653.451	618.952	816813.12	700653.94	0.170799E+11	0.00000
7.57	638.314	610.834	797892.19	691464.69	0.115438E+11	0.00000
8.52	625.258	604.520	781571.25	684316.31	0.583178E+10	0.00000
9.47	613.774	600.000	767216.50	679200.00	0.000000E+00	0.00000

QK-0.597206425E+11
QP,QG,(QG/TAUL)   288.119873   13969956.0   431.811767

```
      66.00 TAUH IN H.MIN

    X         TK        TG          UK              UG          ESP          SA
  0.00     780.000   737.743   1243470.00       835125.75   0.447246E+11   0.96226
  0.95     780.000   725.010   1217843.00       820711.25   0.411409E+11   0.87041
  1.89     780.000   708.439   1180877.00       801952.94   0.374982E+11   0.73792
  2.84     780.000   686.875   1127240.00       777542.12   0.337857E+11   0.54567
  3.79     780.000   658.812   1003210.50       745775.00   0.299853E+11   0.10112
  4.73     691.552   635.616    864438.50       719517.25   0.260438E+11   0.00000
  5.68     660.417   623.448    825520.31       705743.06   0.216879E+11   0.00000
  6.63     643.173   614.903    803964.87       696070.25   0.167189E+11   0.00000
  7.57     630.228   608.333    787783.37       688633.00   0.113355E+11   0.00000
  8.52     619.271   603.385    774087.12       683032.06   0.574077E+10   0.00000
  9.47     609.964   600.000    762454.50       679200.00   0.000000E+00   0.00000
QK-0.597206425E+11
QP,QG,(QG/TAUL)   280.665527         14990747.0        417.042968


      67.00 TAUH IN H.MIN

    X         TK        TG          UK              UG          ESP          SA
  0.00     780.000   733.601   1225387.00       830436.12   0.437264E+11   0.89745
  0.95     780.000   719.619   1194310.00       814608.56   0.401641E+11   0.78606
  1.89     780.000   701.421   1150253.00       794008.69   0.365535E+11   0.62815
  2.84     780.000   677.740   1087354.00       767202.31   0.328843E+11   0.40271
  3.79     762.071   649.620    952587.37       735370.12   0.291405E+11   0.00000
  4.73     675.168   628.823    843958.87       711828.19   0.252789E+11   0.00000
  5.68     649.388   618.739    811734.31       700412.50   0.210888E+11   0.00000
  6.63     634.769   611.704    793460.06       692448.81   0.163310E+11   0.00000
  7.57     623.796   606.405    779743.94       686450.94   0.111121E+11   0.00000
  8.52     614.689   602.536    768360.94       682070.94   0.564369E+10   0.00000
  9.47     607.213   600.000    759015.44       679200.00   0.000000E+00   0.00000
QK-0.597206425E+11
QP,QG,(QG/TAUL)   272.224121         15987679.0        404.355224


      68.00 TAUH IN H.MIN

    X         TK        TG          UK              UG          ESP          SA
  0.00     780.000   728.477   1205189.00       824636.25   0.427660E+11   0.82506
  0.95     780.000   712.951   1168025.00       807060.44   0.392197E+11   0.69185
  1.89     780.000   692.745   1116047.00       784187.44   0.356341E+11   0.50555
  2.84     780.000   666.451   1042557.12       754422.37   0.319997E+11   0.24215
  3.79     728.809   639.941    911010.69       724413.94   0.283025E+11   0.00000
  4.73     661.603   623.284    827002.50       705557.69   0.245071E+11   0.00000
  5.68     640.261   614.950    800324.94       696123.87   0.204535E+11   0.00000
  6.63     627.926   609.180    784906.69       689592.19   0.158972E+11   0.00000
  7.57     618.696   604.921    773369.06       684770.62   0.108626E+11   0.00000
  8.52     611.189   601.901    763985.00       681351.56   0.553526E+10   0.00000
  9.47     605.225   600.000    756530.31       679200.00   0.000000E+00   0.00000
QK-0.597206425E+11
QP,QG,(QG/TAUL)   261.784423         16942176.0        392.797607
```

```
      69.00 TAUH IN H.MIN

    X       TK        TG        UK           UG          ESP          SA
  0.00   780.000   724.387  1183092.00    820006.81  0.418387E+11  0.74585
  0.95   780.000   707.629  1139270.00    801036.75  0.383044E+11  0.58879
  1.89   780.000   685.821  1078625.00    776349.12  0.347384E+11  0.37142
  2.84   780.000   657.440   993338.56    744222.75  0.311320E+11  0.06573
  3.79   702.700   632.152   878373.81    715595.81  0.274729E+11  0.00000
  4.73   650.372   618.774   812963.56    700451.94  0.237322E+11  0.00000
  5.68   632.732   611.907   790914.31    692679.50  0.197875E+11  0.00000
  6.63   622.373   607.192   777965.12    687341.37  0.154221E+11  0.00000
  7.57   614.661   603.778   768325.44    683477.00  0.105896E+11  0.00000
  8.52   608.515   601.424   760643.06    680812.50  0.541663E+10  0.00000
  9.47   603.787   600.000   754732.62    679200.00  0.000000E+00  0.00000
QK-0.597206425E+11
QP,QG,(QG/TAUL)   253.451553       17853312.0        382.076416


      70.00 TAUH IN H.MIN

    X       TK        TG        UK           UG          ESP          SA
  0.00   780.000   718.106  1159088.00    812895.69  0.409461E+11  0.65982
  0.95   780.000   699.453  1108032.00    791780.87  0.374208E+11  0.47682
  1.89   780.000   675.179  1037907.62    764302.50  0.338700E+11  0.22548
  2.84   755.019   647.350   943772.87    732800.19  0.302859E+11  0.00000
  3.79   682.123   625.883   852652.50    708499.56  0.266579E+11  0.00000
  4.73   641.108   615.113   801384.69    696307.44  0.229622E+11  0.00000
  5.68   626.550   609.471   783186.44    689921.25  0.191045E+11  0.00000
  6.63   617.885   605.629   772355.69    685571.87  0.149147E+11  0.00000
  7.57   611.479   602.900   764348.25    682482.87  0.102928E+11  0.00000
  8.52   606.477   601.068   758095.56    680409.00  0.528776E+10  0.00000
  9.47   602.747   600.000   753432.69    679200.00  0.000000E+00  0.00000
QK-0.597206425E+11
QP,QG,(QG/TAUL)   240.651931       18725328.0        372.121826


      71.00 TAUH IN H.MIN

    X       TK        TG        UK           UG          ESP          SA
  0.00   780.000   711.632  1132220.00    805568.00  0.401045E+11  0.56352
  0.95   780.000   691.031  1073068.00    782246.94  0.365859E+11  0.35150
  1.89   780.000   664.219   992073.69    751895.87  0.330469E+11  0.06120
  2.84   723.227   637.879   904032.87    722079.12  0.294807E+11  0.00000
  3.79   665.718   620.821   832146.94    702769.31  0.258781E+11  0.00000
  4.73   633.485   612.146   791855.00    692949.44  0.222193E+11  0.00000
  5.68   621.487   607.522   776857.50    687715.31  0.184317E+11  0.00000
  6.63   614.268   604.401   767834.06    684182.12  0.143824E+11  0.00000
  7.57   608.974   602.225   761217.19    681718.62  0.996010E+10  0.00000
  8.52   604.924   600.801   756154.25    680106.75  0.514352E+10  0.00000
  9.47   601.994   600.000   752491.44    679200.00  0.000000E+00  0.00000
QK-0.597206425E+11
QP,QG,(QG/TAUL)   227.461746       19550384.0        362.623046


      72.00 TAUH IN H.MIN

    X       TK        TG        UK           UG          ESP          SA
  0.00   780.000   702.721  1102457.00    795480.44  0.393146E+11  0.45684
  0.95   780.000   679.431  1034271.44    769116.25  0.358010E+11  0.21245
  1.89   756.014   652.735   945016.31    738895.87  0.322712E+11  0.00000
  2.84   698.174   630.321   872716.00    713523.19  0.287192E+11  0.00000
  3.79   652.615   616.735   815767.81    698144.56  0.251371E+11  0.00000
  4.73   627.225   609.747   784030.69    690233.69  0.215087E+11  0.00000
  5.68   617.353   605.967   771689.62    685954.50  0.177775E+11  0.00000
  6.63   611.360   603.438   764199.50    683091.69  0.138394E+11  0.00000
  7.57   607.006   601.706   758757.19    681131.69  0.959991E+10  0.00000
  8.52   603.741   600.601   754676.06    679880.25  0.498358E+10  0.00000
  9.47   601.448   600.000   751809.56    679200.00  0.000000E+00  0.00000
QK-0.597206425E+11
QP,QG,(QG/TAUL)   209.304428       20322032.0        353.372558
```

NUMERICAL OUTPUT   No. III

```
        11         =M NUMBER OF CALCULATION POINTS IN X-DIRECTION
  1000.00000       =TGEBE [C] INLET TEMPERATURE FLUID (CHARGING)
   600.000000      =TGEEN [C] INLET TEMPERATURE FLUID (DISCHARGING)
   880.000000      =TKS [C] MELTING POINT
   780.000000      =TKK [C] CRYSTALLIZATION POINT
    20.0000000     =TU [C] AMBIENT TEMPERATUR
  2500.00000       =RHOK [KG/M**3] DENSITY SOLID MEDIUM
  1100.00000       =CKK [J/(KG*K)] SPECIFIC HEAT CERAMIC
  1433.33007       =CKS [J/(KG*K)] SPECIFIC HEAT PHASE-CHANGE-MATERIAL
     5.00000000    =RLAMK [W/(M*K)] COEFF. OF THERMAL CONDUCT. SOLID MED.
620000.000         =RK [J/KG] HEAT OF FUSION
     0.449999988   =RMS MASS-SHARE OF SALT
     0.310500025   =RHOGB [KG/M**3] DENSITY FLUID (CHARGING)
     3.66800022    =RHOGE [KG/M**3] DENSITY FLUID (DISCHARGING)
  1163.00000       =CGB [J/(KG*K)] SPECIFIC HEAT FLUID (CHARGING)
  1132.00000       =CGE [J/(KG*K)] SPECIFIC HEAT FLUID (DISCHARGING)
     0.727999806E-01 =RLAMGB [W/(M*K)] COEFF. OF THERMAL COND. FLUID CHARG.
     0.665000081E-01 =RLAMGE [W/(M*K)] COEFF. OF THERM. COND. FLUID DISCH.
     0.709999978   =PRGB PRANDTL-NUMBER (CHARGING)
    69.0000000     =PRGE PRANDTL-NUMBER (DISCHARGING)
   800.000000      =TKVA NOMINAL TEMPERATURE FOR HEAT LOSS
     1.00000000    =DTKVU TEMPERATURE GRADIENT FOR HEAT LOSS
300000.000         =QPDIS[W] ENERGY OUTPUT (DISCHARGING)
 57600.0000        =TAUDIS [S] DISCHARGING-TIME
 28800.0000        =TAUCHA [S] CHARGING-TIME
     0.699999988   =ETASTO EFFICIENCY OF STORING (STARTVALUE)
   700.000000      =TCOUT [C] OUTLET-TEMPERATURE FLUID (CHARGING)
    50.000000      =TDOUT [C] OUTLET-TEMPERATURE FLUID (DISCHARGING)
     0.800000000   =FF FILLING FACTOR
     0.143700002E-03 =RNUEB [M**2/S] VISCOSITY FLUID (CHARGING)
     0.110500003E-02 =RNUEE [M**2/S] VISCOSITY FLUID (DISCHARGING)
   200.000000      =DELTAT [C] AVERAGE TEMPERATURE DIFFERENCE CERAMIC


AB= 0.610534966
AE= 0.147520124
B = 0.199999986E-02
C= 0.113000009E-07
DB=   19654.9570
DE=    402.017578
EB= 0.234460413
EE= 0.181297697E-01
EST  0.246857113E+11
EMST   46664.8984
VS   18.6659545
QPCHA -0.230789184E-03
RMPGB   2.50000000
RMPGE   1.79999923
H    9.05600547
DA   1.81120109
DKRIT  0.213362723E-01
DI  0.219999961E-01
RET   2392.16577
REL   13.9605255
RNUT   10.8183126
RNUL   2.72416591
ALPHAT    35.7987670
ALPHAL    8.23441219
ALPHAB    33.5657348
ALPHAE    8.11030006
NBOHR         1356
AG  0.515459537
AK   2.06099987
U   93.7199554
S   0.557499118E-01
```

```
TAU=     3601.7   TOUT=    680.91   SUMMET=  0.234319E+07
TAU=     7203.5   TOUT=    730.63   SUMMET=  0.485511E+07
TAU=    10806.0   TOUT=    775.91   SUMMET=  0.753592E+07
TAU=    14407.9   TOUT=    814.97   SUMMET=  0.103742E+08
TAU=    18009.6   TOUT=    844.88   SUMMET=  0.133318E+08
TAU=    21611.4   TOUT=    864.59   SUMMET=  0.163839E+08
TAU=    25213.1   TOUT=    875.13   SUMMET=  0.193816E+08
TAU=    28514.7   TOUT=    879.70   SUMMET=  0.221151E+08
***************************************************************
DISCHARGING   TAU=   28800.
***************************************************************
TAU=    32406.2   TOUT=    938.32   SUMMET=  0.343910E+07
TAU=    36013.3   TOUT=    906.28   SUMMET=  0.676411E+07
TAU=    39620.4   TOUT=    875.08   SUMMET=  0.997446E+07
TAU=    43227.4   TOUT=    846.78   SUMMET=  0.130775E+08
TAU=    46834.5   TOUT=    820.84   SUMMET=  0.160816E+08
TAU=    50441.6   TOUT=    799.60   SUMMET=  0.189780E+08
TAU=    54048.7   TOUT=    783.58   SUMMET=  0.218005E+08
TAU=    57655.8   TOUT=    770.31   SUMMET=  0.245740E+08
TAU=    61262.9   TOUT=    761.85   SUMMET=  0.273042E+08
TAU=    64870.0   TOUT=    757.86   SUMMET=  0.300345E+08
TAU=    68477.0   TOUT=    753.21   SUMMET=  0.327152E+08
TAU=    72084.1   TOUT=    750.03   SUMMET=  0.353848E+08
TAU=    75691.2   TOUT=    743.90   SUMMET=  0.380543E+08
TAU=    79298.3   TOUT=    739.14   SUMMET=  0.407103E+08
TAU=    82905.4   TOUT=    730.26   SUMMET=  0.433192E+08
TAU=    86211.9   TOUT=    723.98   SUMMET=  0.457107E+08
***************************************************************
CHARGING      TAU=   86400.
***************************************************************
TAU=    90001.9   TOUT=    620.21   SUMMET=  0.217574E+07
TAU=    93603.6   TOUT=    658.62   SUMMET=  0.444424E+07
TAU=    97205.4   TOUT=    710.54   SUMMET=  0.687663E+07
TAU=   100807.1   TOUT=    763.93   SUMMET=  0.950278E+07
TAU=   104408.9   TOUT=    808.95   SUMMET=  0.123062E+08
TAU=   108010.6   TOUT=    841.30   SUMMET=  0.152520E+08
TAU=   111612.3   TOUT=    861.75   SUMMET=  0.182556E+08
TAU=   114913.9   TOUT=    872.28   SUMMET=  0.209891E+08
***************************************************************
DISCHARGING   TAU=  115200.
***************************************************************
TAU=   118806.2   TOUT=    924.75   SUMMET=  0.339250E+07
TAU=   122413.3   TOUT=    892.26   SUMMET=  0.666758E+07
TAU=   126020.4   TOUT=    860.69   SUMMET=  0.982680E+07
TAU=   129627.4   TOUT=    832.51   SUMMET=  0.128773E+08
TAU=   133234.5   TOUT=    806.99   SUMMET=  0.158310E+08
TAU=   136841.6   TOUT=    786.44   SUMMET=  0.186841E+08
TAU=   140448.7   TOUT=    770.81   SUMMET=  0.214642E+08
TAU=   144055.8   TOUT=    758.83   SUMMET=  0.241944E+08
TAU=   147662.9   TOUT=    752.38   SUMMET=  0.268756E+08
TAU=   151270.0   TOUT=    749.16   SUMMET=  0.295452E+08
TAU=   154877.0   TOUT=    744.93   SUMMET=  0.322147E+08
TAU=   158484.1   TOUT=    739.96   SUMMET=  0.348824E+08
TAU=   162091.2   TOUT=    733.14   SUMMET=  0.374913E+08
TAU=   165698.3   TOUT=    725.64   SUMMET=  0.401002E+08
TAU=   169305.4   TOUT=    715.47   SUMMET=  0.426719E+08
TAU=   172611.9   TOUT=    705.58   SUMMET=  0.450023E+08
***************************************************************
CHARGING      TAU=  172800.
***************************************************************
TAU=   176402.0   TOUT=    615.21   SUMMET=  0.217367E+07
TAU=   180003.7   TOUT=    647.63   SUMMET=  0.441033E+07
TAU=   183605.4   TOUT=    696.16   SUMMET=  0.679543E+07
TAU=   187207.2   TOUT=    749.72   SUMMET=  0.936868E+07
TAU=   190808.9   TOUT=    797.30   SUMMET=  0.121257E+08
TAU=   194410.7   TOUT=    833.10   SUMMET=  0.150354E+08
TAU=   198012.4   TOUT=    856.61   SUMMET=  0.180253E+08
TAU=   201314.0   TOUT=    869.32   SUMMET=  0.207587E+08
***************************************************************
DISCHARGING   TAU=  201600.
***************************************************************
TAU=   205206.2   TOUT=    920.40   SUMMET=  0.337714E+07
TAU=   208813.3   TOUT=    887.98   SUMMET=  0.663658E+07
TAU=   212420.3   TOUT=    856.55   SUMMET=  0.978064E+07
TAU=   216027.4   TOUT=    828.28   SUMMET=  0.128166E+08
TAU=   219634.5   TOUT=    803.78   SUMMET=  0.157563E+08
TAU=   223241.6   TOUT=    783.86   SUMMET=  0.185976E+08
TAU=   226848.7   TOUT=    767.08   SUMMET=  0.213627E+08
TAU=   230455.8   TOUT=    756.19   SUMMET=  0.240884E+08
TAU=   234062.9   TOUT=    751.02   SUMMET=  0.267580E+08
TAU=   237670.0   TOUT=    748.47   SUMMET=  0.294276E+08
TAU=   241277.0   TOUT=    743.20   SUMMET=  0.320971E+08
TAU=   244884.1   TOUT=    738.70   SUMMET=  0.347460E+08
TAU=   248491.2   TOUT=    730.75   SUMMET=  0.373548E+08
TAU=   252098.3   TOUT=    723.86   SUMMET=  0.399637E+08
TAU=   255705.4   TOUT=    712.21   SUMMET=  0.425181E+08
TAU=   259011.9   TOUT=    703.06   SUMMET=  0.448317E+08
```

```
      48.00 TAUH IN H.MIN

     X       TK        TG         UK          UG         ESP        SA
  0.00    780.000   795.095   1089858.00   798167.62   0.384759E+11  0.41168
  0.91    780.000   674.645    998756.87   763697.81   0.349714E+11  0.08516
  1.81    719.011   644.211    898763.25   729246.87   0.314589E+11  0.00000
  2.72    668.566   624.053    835706.19   706427.81   0.279335E+11  0.00000
  3.62    637.071   612.358    796337.25   693189.62   0.243890E+11  0.00000
  4.53    617.030   606.385    771286.75   686427.44   0.208115E+11  0.00000
  5.43    609.686   603.550    762107.00   683218.25   0.171542E+11  0.00000
  6.34    605.734   601.858    757166.44   681303.12   0.133465E+11  0.00000
  7.24    603.098   600.818    753871.94   680125.94   0.929993E+10  0.00000
  8.15    601.354   600.245    751692.12   679477.87   0.487291E+10  0.00000
  9.06    600.344   600.000    750429.75   679200.00   0.000000E+00  0.00000
QK-0.606754529E+11
QP,QG,(QG/TAUL)   214.141372          22001632.0          382.733398


*******************************************************************
CHARGING    TAU=   172800.
*******************************************************************


      49.00 TAUH IN H.MIN

     X       TK        TG         UK          UG         ESP        SA
  0.00    886.543  1000.000   1387178.00  1163000.00   0.425866E+11  1.00000
  0.71    880.000   923.232   1135041.00  1073719.00   0.367019E+11  0.12560
  1.81    830.836   878.635   1038543.25  1021853.00   0.316308E+11  0.00000
  2.72    773.651   828.397    967062.56   963425.44   0.269517E+11  0.00000
  3.62    723.625   775.945    904530.25   902424.37   0.225851E+11  0.00000
  4.53    683.338   729.284    854171.56   846995.00   0.184820E+11  0.00000
  5.43    654.342   689.197    817926.37   801536.37   0.145809E+11  0.00000
  6.34    634.544   659.755    793178.81   767295.06   0.108221E+11  0.00000
  7.24    621.441   638.881    776800.44   743019.19   0.715930E+10  0.00000
  8.15    612.952   624.604    766189.06   726414.50   0.355942E+10  0.00000
  9.06    607.572   615.188    759463.50   715464.06   0.000000E+00  0.00000
QK 0.425865625E+11
QP,QG,(QG/TAUL)  -1118.83984         -4086866.00         -1137.73706


      50.00 TAUH IN H.MIN
 179999.004552125930
     X       TK        TG         UK          UG         ESP        SA
  0.00    980.190  1000.000   1504236.00  1163000.00   0.464636E+11  1.00000
  0.91    880.000   954.027   1273811.00  1109533.00   0.399821E+11  0.62299
  1.81    880.000   905.344   1111205.00  1052915.00   0.344176E+11  0.04017
  2.72    849.583   878.676   1061978.00  1021900.00   0.293475E+11  0.00000
  3.62    806.555   845.395   1008192.37   983194.06   0.245177E+11  0.00000
  4.53    763.076   805.552    953844.12   936857.19   0.199402E+11  0.00000
  5.43    723.591   764.637    904487.25   889272.62   0.156046E+11  0.00000
  6.34    690.407   726.735    863007.94   845192.56   0.114809E+11  0.00000
  7.24    664.112   694.185    830139.50   807283.69   0.753072E+10  0.00000
  8.15    644.213   667.877    805264.75   776741.62   0.371523E+10  0.00000
  9.06    629.738,  647.556    787171.06   753107.31   0.000000E+00  0.00000
QK 0.464636477E+11
QP,QG,(QG/TAUL)  -1024.73168         -7930894.00         -1130.48095
```

```
    51.00 TAUH IN H.MIN

    X        TK       TG         UK          UG          ESP         SA
  0.00    996.263  1000.000  1524327.00  1163000.00  0.499104E+11  1.00000
  0.91    922.448   973.269  1432058.00  1131912.00  0.430127E+11  1.00000
  1.81    880.000   925.887  1177764.00  1076807.00  0.369237E+11  0.27873
  2.72    880.000   895.717  1102554.00  1041718.62  0.316036E+11  0.00916
  3.62    857.081   877.826  1071350.00  1020912.19  0.265318E+11  0.00000
  4.53    825.367   853.789  1031707.62   992956.62  0.216252E+11  0.00000
  5.43    790.134   823.470   987666.25   957695.62  0.169139E+11  0.00000
  6.34    754.793   789.946   943489.75   918706.94  0.124084E+11  0.00000
  7.24    721.975   756.062   902467.50   879300.62  0.810172E+10  0.00000
  8.15    693.294   724.166   866617.06   842205.50  0.397436E+10  0.00000
  9.06    669.509   696.046   836885.37   809501.44  0.000000E+00  0.00000
QK  0.499104194E+11
QP,QG,(QG/TAUL) -883.746337        -11345135.0       -1089.69751


    52.00 TAUH IN H.MIN

    X        TK       TG         UK          UG          ESP         SA
  0.00    999.232  1000.000  1528038.00  1163000.00  0.528112E+11  1.00000
  0.91    975.469   991.680  1498334.00  1153324.00  0.457503E+11  1.00000
  1.81    880.000   949.626  1310328.00  1104415.00  0.391973E+11  0.75387
  2.72    880.000   903.837  1147635.00  1051163.00  0.334626E+11  0.17074
  3.62    880.000   888.161  1100530.00  1032931.81  0.282175E+11  0.00191
  4.53    862.477   877.032  1078095.00  1019988.75  0.231347E+11  0.00000
  5.43    838.212   859.481  1047764.37   999576.81  0.181749E+11  0.00000
  6.34    809.592   836.081  1011988.75   972361.75  0.133694E+11  0.00000
  7.24    778.939   808.588   973673.12   940387.62  0.873677E+10  0.00000
  8.15    748.450   779.053   935561.94   906039.00  0.428243E+10  0.00000
  9.06    719.989   749.584   899985.25   871766.44  0.000000E+00  0.00000
QK  0.528112435E+11
QP,QG,(QG/TAUL) -728.084472        -14214485.0       -1029.51025


    53.00 TAUH IN H.MIN

    X        TK       TG         UK          UG          ESP         SA
  0.00    999.828  1000.000  1528783.00  1163000.00  0.551747E+11  1.00000
  0.91    992.056   997.330  1519068.00  1159895.00  0.480636E+11  1.00000
  1.81    935.271   975.188  1448087.00  1134144.00  0.411408E+11  1.00000
  2.72    880.000   930.760  1220271.00  1082474.00  0.349152E+11  0.43107
  3.62    880.000   897.380  1125274.00  1043653.62  0.294429E+11  0.09059
  4.53    878.927   885.597  1098658.00  1029949.31  0.242543E+11  0.00000
  5.43    865.566   876.818  1081956.00  1019739.62  0.191668E+11  0.00000
  6.34    846.841   863.259  1058550.00  1003970.00  0.141729E+11  0.00000
  7.24    823.700   844.858  1029623.44   982570.44  0.930110E+10  0.00000
  8.15    797.618   822.359   997021.62   956403.56  0.457284E+10  0.00000
  9.06    770.403   797.149   963002.56   927084.69  0.000000E+00  0.00000
QK  0.551746969E+11
QP,QG,(QG/TAUL) -589.788574        -16547131.0       -961.885742
```

```
      54.00 TAUH IN H.MIN

     X        TK        TG        UK           UG          ESP          SA
   0.00    999.957  1000.000  1528945.00   1163000.00  0.570987E+11  1.00000
   0.91    997.420   999.136  1525773.00   1161996.00  0.499716E+11  1.00000
   1.81    977.937   991.600  1501419.00   1153231.00  0.429088E+11  1.00000
   2.72    880.000   950.411  1358010.00   1105328.00  0.362373E+11  0.92477
   3.62    880.000   904.106  1172188.00   1051475.00  0.303341E+11  0.25875
   4.53    880.000   888.253  1114734.00   1033038.50  0.249985E+11  0.05282
   5.43    878.179   882.227  1097723.00   1026030.50  0.198367E+11  0.00000
   6.34    868.026   876.226  1085031.00   1019050.81  0.147443E+11  0.00000
   7.24    853.270   865.984  1066586.00   1007139.50  0.972444E+10  0.00000
   8.15    834.392   851.409  1042988.87    990188.87  0.480270E+10  0.00000
   9.06    812.448   833.014  1015558.12    968795.00  0.000000E+00  0.00000
 QK 0.570987356E+11
 QP,QG,(QG/TAUL) -485.512451        -17870688.0        -867.569335


      55.00 TAUH IN H.MIN

     X        TK        TG        UK           UG          ESP          SA
   0.00    999.989  1000.000  1528984.00   1163000.00  0.587133E+11  1.00000
   0.91    999.160   999.719  1527949.00   1162674.00  0.515810E+11  1.00000
   1.81    992.365   997.117  1519455.00   1159647.00  0.444710E+11  1.00000
   2.72    951.822   980.660  1468776.00   1140508.00  0.374990E+11  1.00000
   3.62    880.000   938.077  1263169.00   1090984.00  0.311250E+11  0.58484
   4.53    880.000   899.883  1145772.00   1046563.56  0.255047E+11  0.16406
   5.43    880.000   886.810  1108673.00   1031359.62  0.202450E+11  0.03109
   6.34    878.091   881.703  1097612.00   1025421.19  0.150976E+11  0.00000
   7.24    869.965   876.646  1087455.00   1019539.75  0.999972E+10  0.00000
   8.15    858.102   868.375  1072626.00   1009920.00  0.496014E+10  0.00000
   9.06    842.727   856.531  1053407.00    996145.44  0.000000E+00  0.00000
 QK 0.587132518E+11
 QP,QG,(QG/TAUL) -417.136718        -18864192.0        -786.193603


      56.00 TAUH IN H.MIN

     X        TK        TG        UK           UG          ESP          SA
   0.00    999.996  1000.000  1528994.00   1163000.00  0.601362E+11  1.00000
   0.91    999.725   999.908  1528655.00   1162893.00  0.530023E+11  1.00000
   1.81    997.368   999.012  1525709.00   1161851.00  0.458760E+11  1.00000
   2.72    982.923   993.182  1507652.00   1155071.00  0.387987E+11  1.00000
   3.62    902.337   959.930  1406920.00   1116399.00  0.319986E+11  1.00000
   4.53    880.000   914.726  1198118.00   1063826.00  0.259207E+11  0.35168
   5.43    880.000   891.857  1126585.00   1037229.75  0.204970E+11  0.09529
   6.34    880.000   884.104  1104176.00   1028213.31  0.152925E+11  0.01497
   7.24    877.842   880.658  1097301.00   1024205.37  0.101564E+11  0.00000
   8.15    871.417   876.712  1089270.00   1019616.56  0.505499E+10  0.00000
   9.06    861.936   870.087  1077419.00   1011911.62  0.000000E+00  0.00000
 QK 0.601361940E+11
 QP,QG,(QG/TAUL) -377.721435        -19857712.0        -724.994384
```

```
********************************************************
DISCHARGING   TAU=   201600.
********************************************************


   57.00 TAUH IN H.MIN

    X       TK       TG        UK          UG          ESP          SA
  0.00   977.657   920.460  1501070.00  1041961.19  0.576682E+11  1.00000
  0.91   970.345   898.689  1491930.00  1017316.62  0.530868E+11  1.00000
  1.81   959.511   871.752  1478387.00   986823.87  0.483600E+11  1.00000
  2.72   938.620   840.316  1452274.00   951237.44  0.435158E+11  1.00000
  3.62   876.963   812.877  1375203.00   920176.50  0.385344E+11  1.00000
  4.53   855.151   791.250  1167078.00   895695.56  0.332793E+11  0.35176
  5.43   847.017   766.919  1085364.00   868152.00  0.273479E+11  0.09532
  6.34   836.450   736.496  1049741.00   833713.69  0.207509E+11  0.01498
  7.24   821.288   698.936  1026608.50   791196.12  0.139133E+11  0.00000
  8.15   799.544   653.611   999428.31   739887.37  0.698310E+10  0.00000
  9.06   771.407   600.000   964257.25   679200.00  0.000000E+00  0.00000
QK-0.601365422E+11
QP,QG,(QG/TAUL)   652.969238        2468639.00        685.875732


   58.00 TAUH IN H.MIN

    X       TK       TG        UK          UG          ESP          SA
  0.00   951.788   888.101  1468734.00  1005330.31  0.554364E+11  1.00000
  0.91   939.189   864.764  1452985.00   978912.75  0.511843E+11  1.00000
  1.81   922.737   837.845  1432420.00   948440.25  0.467527E+11  1.00000
  2.72   898.413   808.270  1402015.00   914961.94  0.421655E+11  1.00000
  3.62   849.110   781.639  1340386.00   884816.00  0.374020E+11  1.00000
  4.53   826.946   758.710  1131821.00   858859.56  0.323285E+11  0.35176
  5.43   812.966   733.804  1042800.37   830666.69  0.265605E+11  0.09532
  6.34   795.790   705.108   998916.62   798182.06  0.201620E+11  0.01498
  7.24   773.809   672.705   967260.44   761501.62  0.135489E+11  0.00000
  8.15   745.794   637.293   932241.00   721415.37  0.681680E+10  0.00000
  9.06   712.247   600.000   890308.19   679200.00  0.000000E+00  0.00000
QK-0.601365422E+11
QP,QG,(QG/TAUL)   587.033935        4698684.00        653.224365


   59.00 TAUH IN H.MIN

    X       TK       TG        UK          UG          ESP          SA
  0.00   923.815   856.728  1433768.00   969816.25  0.534392E+11  1.00000
  0.91   907.427   832.782  1413282.00   942709.00  0.494160E+11  1.00000
  1.81   887.438   806.491  1388296.00   912947.81  0.452135E+11  1.00000
  2.72   861.580   778.834  1355974.00   881640.12  0.408466E+11  1.00000
  3.62   820.598   753.519  1304746.00   852983.62  0.362851E+11  1.00000
  4.53   798.093   730.818  1095755.00   827285.94  0.313904E+11  0.35176
  5.43   780.473   707.045  1002183.56   800374.81  0.257898E+11  0.09532
  6.34   762.374   680.867   952966.81   770742.00  0.195819E+11  0.00000
  7.24   735.030   653.286   918786.19   739519.87  0.131791E+11  0.00000
  8.15   705.995   625.953   882492.94   708579.25  0.664259E+10  0.00000
  9.06   673.586   600.000   841981.31   679200.00  0.000000E+00  0.00000
QK-0.601365422E+11
QP,QG,(QG/TAUL)   523.108642        6693987.00        620.680908
```

```
      60.00 TAUH IN H.MIN

    X       TK        TG        UK            UG          ESP         SA
 0.00    895.183   828.491  1397977.00    937852.25  0.516611E+11  1.00000
 0.91    876.360   805.200  1374449.00    911486.75  0.477969E+11  1.00000
 1.81    854.501   780.709  1347125.00    883763.00  0.437704E+11  1.00000
 2.72    828.525   755.985  1314655.00    855775.25  0.395872E+11  1.00000
 3.62    793.667   733.577  1271082.00    830409.06  0.352192E+11  1.00000
 4.53    780.000   711.925  1061707.00    805898.81  0.305043E+11  0.31078
 5.43    767.337   686.820   959170.62    777480.00  0.250616E+11  0.00000
 6.34    730.472   661.576   913089.44    748904.19  0.190287E+11  0.00000
 7.24    703.943   638.957   879928.06    723299.00  0.128183E+11  0.00000
 8.15    676.729   618.067   845910.37    699652.00  0.646849E+10  0.00000
 9.06    648.284   600.000   810354.50    679200.00  0.000000E+00  0.00000
QK-0.601365422E+11
QP,QG,(QG/TAUL)   465.573242        8470306.00       589.127929


      61.00 TAUH IN H.MIN

    X       TK        TG        UK            UG          ESP         SA
 0.00    867.271   803.991  1363088.00    910118.19  0.500790E+11  1.00000
 0.91    847.252   782.330  1338064.00    885597.75  0.463255E+11  1.00000
 1.81    824.914   760.474  1310141.00    860857.25  0.424329E+11  1.00000
 2.72    799.972   739.344  1278964.00    836937.25  0.383990E+11  1.00000
 3.62    780.000   718.756  1240369.00    813632.00  0.342007E+11  0.95115
 4.53    780.000   693.859  1020374.69    785448.56  0.296743E+11  .16264
 5.43    735.802   667.822   919751.69    755974.19  0.243997E+11  0.00000
 6.34    703.806   646.685   879756.62    732047.94  0.185217E+11  0.00000
 7.24    679.405   628.422   849254.81    711373.50  0.124809E+11  0.00000
 8.15    655.350   612.583   819186.94    693444.56  0.630220E+10  0.00000
 9.06    631.720   600.000   789648.94    679200.00  0.000000E+00  0.00000
QK-0.601365422E+11
QP,QG,(QG/TAUL)   415.652099        10050601.0       559.312988


      62.00 TAUH IN H.MIN

    X       TK        TG        UK            UG          ESP         SA
 0.00    841.448   784.077  1330808.00    887575.81  0.486580E+11  1.00000
 0.91    821.401   764.827  1305750.00    865784.87  0.449815E+11  1.00000
 1.81    799.828   746.210  1278784.00    844709.44  0.411901E+11  1.00000
 2.72    780.000   728.441  1250175.00    824595.50  0.372761E+11  0.98630
 3.62    780.000   707.481  1205045.00    800868.19  0.332183E+11  0.82454
 4.53    776.159   678.776   970198.00    768374.31  0.288856E+11  0.00000
 5.43    709.534   652.726   886916.94    738886.00  0.238105E+11  0.00000
 6.34    681.864   635.253   852328.81    719106.75  0.180821E+11  0.00000
 7.24    660.259   620.694   825322.12    702626.19  0.121816E+11  0.00000
 8.15    639.804   608.767   799753.75    689124.37  0.615150E+10  0.00000
 9.06    620.859   600.000   776072.94    679200.00  0.000000E+00  0.00000
QK-0.601365422E+11
QP,QG,(QG/TAUL)   375.075683        11469804.0       531.936767
```

```
      63.00 TAUH IN H.MIN

   X       TK        TG        UK          UG        ESP        SA
 0.00    818.248   767.243  1301809.00   868519.62  0.473721E+11  1.00000
 0.91    799.021   750.415  1277775.00   849470.19  0.437493E+11  1.00000
 1.81    780.000   734.522  1252833.00   831478.87  0.400338E+11  0.99582
 2.72    780.000   716.033  1219457.00   810549.94  0.362147E+11  0.87620
 3.62    780.000   690.030  1161524.00   781114.44  0.322723E+11  0.66855
 4.53    739.002   661.786   923751.37   749142.25  0.281114E+11  0.00000
 5.43    687.768   640.807   859709.37   725393.12  0.232463E+11  0.00000
 6.34    664.078   626.529   830096.81   709230.69  0.176911E+11  0.00000
 7.24    645.475   615.043   806843.12   696229.44  0.119229E+11  0.00000
 8.15    628.552   606.111   785689.50   686118.00  0.601858E+10  0.00000
 9.06    613.735   600.000   767167.37   679200.00  0.000000E+00  0.00000
OK-0.601365422E+11
OP,OG,(OG/TAUL)  340.774902        12753896.0         507.028808


      64.00 TAUH IN H.MIN

   X       TK        TG        UK          UG        ESP        SA
 0.00    798.392   756.336  1276989.00   856173.00  0.461869E+11  1.00000
 0.91    780.787   742.817  1254983.00   840869.12  0.426013E+11  1.00000
 1.81    780.000   727.541  1226835.00   823575.94  0.389426E+11  0.90264
 2.72    780.000   706.216  1182774.00   799436.25  0.351981E+11  0.74471
 3.62    780.000   676.222  1109435.00   765483.25  0.313496E+11  0.48185
 4.53    709.753   648.310   887190.81   733887.50  0.273262E+11  0.00000
 5.43    669.843   631.442   837302.12   714792.75  0.226679E+11  0.00000
 6.34    649.824   619.899   812279.62   701725.50  0.173199E+11  0.00000
 7.24    634.149   610.918   792685.06   691559.69  0.116980E+11  0.00000
 8.15    620.434   604.261   775541.94   684024.06  0.590750E+10  0.00000
 9.06    609.054   600.000   761316.12   679200.00  0.000000E+00  0.00000
OK-0.601365422E+11
OP,OG,(OG/TAUL)  318.550781        13937389.0         484.829589


      65.00 TAUH IN H.MIN

   X       TK        TG        UK          UG        ESP        SA
 0.00    782.741   751.095  1257425.00   850239.56  0.450610E+11  1.00000
 0.91    780.000   738.788  1234178.00   836307.75  0.415014E+11  0.92896
 1.81    780.000   722.034  1197461.00   817342.69  0.378848E+11  0.79736
 2.72    780.000   698.472  1141458.00   790670.44  0.341984E+11  0.59663
 3.62    780.000   665.331  1050603.00   753154.37  0.304254E+11  0.27078
 4.53    686.718   637.675   858395.87   721847.81  0.265118E+11  0.00000
 5.43    655.242   624.133   819051.69   706518.87  0.220580E+11  0.00000
 6.34    638.520   614.884   798149.37   696049.25  0.169437E+11  0.00000
 7.24    625.536   607.914   781919.00   688538.94  0.114867E+11  0.00000
 8.15    614.598   602.973   768246.50   682565.81  0.581330E+10  0.00000
 9.06    605.976   600.000   757469.44   679200.00  0.000000E+00  0.00000
OK-0.601365422E+11
OP,OG,(OG/TAUL)  307.870849        15061378.0         465.736816
```

```
         66.00 TAUH IN H.MIN
   X       TK        TG        UK           UG          ESP         SA
 0.00    780.000   748.508  1241125.00   847311.00  0.439632E+11  0.95386
 0.91    780.000   735.707  1211441.00   832819.81  0.404218E+11  0.84746
 1.81    780.000   717.701  1165477.00   812437.75  0.368364E+11  0.68272
 2.72    780.000   692.378  1096473.00   783772.25  0.331950E+11  0.43539
 3.62    780.000   656.760   986430.50   743452.31  0.294823E+11  0.04097
 4.53    668.563   629.311   835702.69   712379.87  0.256560E+11  0.00000
 5.43    643.452   618.456   804314.25   700092.44  0.214049E+11  0.00000
 6.34    629.626   611.104   787031.75   691769.81  0.165453E+11  0.00000
 7.24    619.022   605.729   773776.69   685685.31  0.112679E+11  0.00000
 8.15    610.411   602.075   763012.31   681549.25  0.572221E+10  0.00000
 9.06    603.950   600.000   754936.25   679200.00  0.000000E+00  0.00000
QK-0.601365422E+11
QP,QG,(QG/TAUL)    302.599365         16156980.0        449.659179


         67.00 TAUH IN H.MIN
   X       TK        TG        UK           UG          ESP         SA
 0.00    780.000   743.308  1223131.00   841424.81  0.428915E+11  0.88936
 0.91    780.000   728.393  1186132.00   824540.87  0.393627E+11  0.75675
 1.81    780.000   707.414  1129879.00   800792.25  0.358003E+11  0.55513
 2.72    780.000   677.907  1046257.00   767391.31  0.321934E+11  0.25541
 3.62    740.856   644.357   926069.56   729412.56  0.285285E+11  0.00000
 4.53    654.165   622.747   817705.75   704949.87  0.247717E+11  0.00000
 5.43    634.012   614.070   792514.19   695127.31  0.207034E+11  0.00000
 6.34    622.683   608.265   778352.75   688556.19  0.161019E+11  0.00000
 7.24    614.124   604.143   767653.69   683890.06  0.110247E+11  0.00000
 8.15    607.415   601.449   759267.44   680840.62  0.562116E+10  0.00000
 9.06    602.614   600.000   753266.94   679200.00  0.000000E+00  0.00000
QK-0.601365422E+11
QP,QG,(QG/TAUL)    292.004394         17214608.0        435.554199


         68.00 TAUH IN H.MIN
   X       TK        TG        UK           UG          ESP         SA
 0.00    780.000   738.788  1202385.00   836308.62  0.418574E+11  0.81501
 0.91    780.000   722.036  1156953.00   817344.37  0.383374E+11  0.65217
 1.81    780.000   698.473  1088839.00   790672.06  0.347919E+11  0.40803
 2.72    780.000   665.333   988025.94   753156.50  0.312115E+11  0.04669
 3.62    705.347   633.891   881682.19   717564.44  0.275842E+11  0.00000
 4.53    642.614   617.592   803265.81   699113.87  0.238830E+11  0.00000
 5.43    626.501   610.694   783125.94   691305.56  0.199520E+11  0.00000
 6.34    617.294   606.138   771617.00   686148.56  0.155898E+11  0.00000
 7.24    610.455   602.993   763067.50   682588.19  0.107444E+11  0.00000
 8.15    605.274   601.012   756591.75   680346.19  0.550466E+10  0.00000
 9.06    601.733   600.000   752164.94   679200.00  0.000000E+00  0.00000
QK-0.601365422E+11
QP,QG,(QG/TAUL)    282.794921         18221392.0        422.610595
```

```
      69.00 TAUH IN H.MIN

     X      TK        TG        UK          UG        ESP         SA
   0.00   780.000   730.954  1178587.00   827439.75  0.408655E+11  0.72971
   0.91   780.000   711.016  1123481.00   804870.00  0.373517E+11  0.53219
   1.81   780.000   682.973  1041632.75   773125.62  0.338187E+11  0.23883
   2.72   744.475   650.747   930593.19   736645.69  0.302583E+11  0.00000
   3.62   679.184   625.213   848979.44   708533.12  0.266604E+11  0.00000
   4.53   633.398   613.562   791746.81   694551.81  0.230035E+11  0.00000
   5.43   620.564   608.105   775704.37   688375.50  0.191756E+11  0.00000
   6.34   613.137   604.550   766420.25   684380.44  0.150238E+11  0.00000
   7.24   607.718   602.160   759647.00   681645.56  0.104224E+11  0.00000
   8.15   603.748   600.708   754683.69   680001.19  0.537102E+10  0.00000
   9.06   601.150   600.000   751436.62   679200.00  0.000000E+00  0.00000
QK-0.601365422E+11
QP,QG,(QG/TAUL)   266.830810        19181952.0        410.668945


      70.00 TAUH IN H.MIN

     X      TK        TG        UK          UG        ESP         SA
   0.00   780.000   724.027  1150620.00   819598.56  0.399316E+11  0.62947
   0.91   780.000   701.275  1084144.00   793843.62  0.364221E+11  0.39120
   1.81   780.000   669.273   985833.62   757617.44  0.328982E+11  0.03884
   2.72   709.667   638.554   887082.94   722842.94  0.293530E+11  0.00000
   3.62   659.670   619.805   824586.19   701620.00  0.257781E+11  0.00000
   4.53   626.086   610.425   782606.94   691001.12  0.221561E+11  0.00000
   5.43   615.893   606.129   769871.12   686137.69  0.184064E+11  0.00000
   6.34   609.946   603.367   762431.69   683011.19  0.144130E+11  0.00000
   7.24   605.685   601.558   757105.31   680963.81  0.100434E+11  0.00000
   8.15   602.661   600.495   753325.25   679760.37  0.521396E+10  0.00000
   9.06   600.765   600.000   750954.81   679200.00  0.000000E+00  0.00000
QK-0.601365422E+11
QP,QG,(QG/TAUL)   252.716995        20090112.0        399.398681


      71.00 TAUH IN H.MIN

     X      TK        TG        UK          UG        ESP         SA
   0.00   780.000   712.542  1118028.00   806597.31  0.390613E+11  0.51265
   0.91   780.000   685.119  1038171.44   775554.50  0.355547E+11  0.22643
   1.81   744.433   653.774   930540.06   740072.87  0.320375E+11  0.00000
   2.72   683.407   629.320   854258.00   712390.81  0.285038E+11  0.00000
   3.62   645.052   615.128   806313.56   696324.75  0.249467E+11  0.00000
   4.53   620.308   607.992   775383.94   688246.62  0.213521E+11  0.00000
   5.43   612.246   604.623   765306.81   684433.19  0.176620E+11  0.00000
   6.34   607.507   602.487   759382.31   682005.19  0.137878E+11  0.00000
   7.24   604.178   601.123   755221.56   680471.06  0.962375E+10  0.00000
   8.15   601.888   600.346   752358.81   679592.06  0.503062E+10  0.00000
   9.06   600.509   600.000   750635.44   679200.00  0.000000E+00  0.00000
QK-0.601365422E+11
QP,QG,(QG/TAUL)   229.314926        20924432.0        388.253906


      72.00 TAUH IN H.MIN

     X      TK        TG        UK          UG        ESP         SA
   0.00   780.000   702.613  1079453.00   795357.50  0.382740E+11  0.37439
   0.91   780.000   671.153   983447.50   759745.12  0.347695E+11  0.03028
   1.81   710.596   641.010   888243.81   725623.69  0.312721E+11  0.00000
   2.72   663.446   622.304   829306.12   704447.87  0.277321E+11  0.00000
   3.62   634.073   611.547   792590.87   692271.69  0.241886E+11  0.00000
   4.53   615.763   606.111   769702.56   686117.87  0.206157E+11  0.00000
   5.43   609.404   603.430   761753.44   683139.06  0.169708E+11  0.00000
   6.34   605.650   601.834   757061.56   681276.50  0.131868E+11  0.00000
   7.24   603.065   600.809   753830.00   680115.50  0.917966E+10  0.00000
   8.15   601.338   600.242   751672.25   679474.44  0.481290E+10  0.00000
   9.06   600.340   600.000   750423.50   679200.00  0.000000E+00  0.00000
QK-0.601365422E+11
QP,QG,(QG/TAUL)   209.083175        21680752.0        377.151367
```

NUMERICAL OUTPUT   No. IV

```
        11          =M NUMBER OF CALCULATION POINTS IN X-DIRECTION
   1000.00000       =TGEBE [C] INLET TEMPERATURE FLUID (CHARGING)
    600.000000      =TGEEN [C] INLET TEMPERATURE FLUID (DISCHARGING)
    880.000000      =TKS [C] MELTING POINT
    780.000000      =TKK [C] CRYSTALLIZATION POINT
     20.0000000     =TU [C] AMBIENT TEMPERATUR
   2500.00000       =RHOK [KG/M**3] DENSITY SOLID MEDIUM
   1100.00000       =CKK [J/(KG*K)] SPECIFIC HEAT CERAMIC
   1433.33007       =CKS [J/(KG*K)] SPECIFIC HEAT PHASE-CHANGE-MATERIAL
      5.00000000    =RLAMK [W/(M*K)] COEFF. OF THERMAL CONDUCT. SOLID MED.
    620000.000      =RK [J/KG] HEAT OF FUSION
      0.000000000   =RMS MASS-SHARE OF SALT
      0.310500025   =RHOGB [KG/M**3] DENSITY FLUID (CHARGING)
      3.66800022    =RHOGE [KG/M**3] DENSITY FLUID (DISCHARGING)
   1163.00000       =CGB [J/(KG*K)] SPECIFIC HEAT FLUID (CHARGING)
   1132.00000       =CGE [J/(KG*K)] SPECIFIC HEAT FLUID (DISCHARGING)
      0.727999806E-01 =RLAMGB [W/(M*K)] COEFF. OF THERMAL COND. FLUID CHARG.
      0.665000081E-01 =RLAMGE [W/(M*K)] COEFF. OF THERM. COND. FLUID DISCH.
      0.709999978   =PRGB PRANDTL-NUMBER (CHARGING)
     69.0000000     =PRGE PRANDTL-NUMBER (DISCHARGING)
    800.000000      =TKVA NOMINAL TEMPERATURE FOR HEAT LOSS
      1.00000000    =DTKVU TEMPERATURE GRADIENT FOR HEAT LOSS
    300000.000      =QPDIS[W] ENERGY OUTPUT (DISCHARGING)
     57600.0000     =TAUDIS [S] DISCHARGING-TIME
     28800.0000     =TAUCHA [S] CHARGING-TIME
      0.699999988   =ETASTO EFFICIENCY OF STORING (STARTVALUE)
    700.000000      =TCOUT [C] OUTLET-TEMPERATURE FLUID (CHARGING)
    750.000000      =TDOUT [C] OUTLET-TEMPERATURE FLUID (DISCHARGING)
      0.700000000   =FF FILLING FACTOR
      0.143700002E-03 =RNUEB [M**2/S] VISCOSITY FLUID (CHARGING)
      0.110500003E-02 =RNUEE [M**2/S] VISCOSITY FLUID (DISCHARGING)
    200.000000      =DELTAT [C] AVERAGE TEMPERATURE DIFFERENCE CERAMIC


AB= 0.135861516
AE= 0.403496133E-01
B= 0.172413792E-02
C= 0.118000009E-07
DB= 2955.23193
DE= 74.2962036
EB= 0.234460413
EE= 0.181097697E-01
EST 0.246857113E+11
EMST  112207.750
VS  38.6923217
QPCHA -0.230789134E-03
RMPGB  2.50000000
RMPGE  1.79999923
H  12.0723629
DA  2.41447358
DKRIT 0.568750351E-01
DI 0.569999963E-01
RET  2325.09838
REL  18.4289245
RNUT  10.5749750
RNUL  3.36043071
ALPHAT  13.5062799
ALPHAL  3.72050266
ALPHAB  13.0757942
ALPHAE  3.88339042
NBOHR       539
AG  1.37539768
AK  3.20322132
U  96.5191345
S  0.731247663E-01
```

```
TAU=    3600.7   TOUT=   678.82   SUMMET= 0.239569E+07
TAU=    7201.4   TOUT=   702.38   SUMMET= 0.487392E+07
TAU=   10802.1   TOUT=   725.64   SUMMET= 0.743650E+07
TAU=   14402.8   TOUT=   748.33   SUMMET= 0.100816E+08
TAU=   18003.5   TOUT=   770.35   SUMMET= 0.128070E+08
TAU=   21604.2   TOUT=   791.66   SUMMET= 0.156110E+08
TAU=   25204.9   TOUT=   812.06   SUMMET= 0.184302E+08
TAU=   28505.5   TOUT=   829.86   SUMMET= 0.209910E+08
************************************************************
DISCHARGING   TAU=   28800.
************************************************************
TAU=   32410.5   TOUT=   884.69   SUMMET= 0.322185E+07
TAU=   36024.2   TOUT=   870.97   SUMMET= 0.639336E+07
TAU=   39637.8   TOUT=   857.63   SUMMET= 0.951599E+07
TAU=   43251.5   TOUT=   844.69   SUMMET= 0.125911E+08
TAU=   46865.2   TOUT=   832.17   SUMMET= 0.156203E+08
TAU=   50478.8   TOUT=   820.08   SUMMET= 0.185992E+08
TAU=   54092.5   TOUT=   808.42   SUMMET= 0.215337E+08
TAU=   57706.2   TOUT=   797.20   SUMMET= 0.244253E+08
TAU=   61319.9   TOUT=   786.42   SUMMET= 0.272781E+08
TAU=   64933.5   TOUT=   776.08   SUMMET= 0.300930E+08
TAU=   68547.2   TOUT=   766.18   SUMMET= 0.328713E+08
TAU=   72160.9   TOUT=   756.72   SUMMET= 0.356142E+08
TAU=   75774.6   TOUT=   747.68   SUMMET= 0.383234E+08
TAU=   79388.2   TOUT=   739.06   SUMMET= 0.410005E+08
TAU=   83001.9   TOUT=   730.85   SUMMET= 0.436475E+08
TAU=   86314.4   TOUT=   723.67   SUMMET= 0.460497E+08
************************************************************
CHARGING   TAU=   86402.
************************************************************
TAU=   90002.9   TOUT=   670.70   SUMMET= 0.237828E+07
TAU=   93603.6   TOUT=   689.46   SUMMET= 0.481768E+07
TAU=   97204.3   TOUT=   709.43   SUMMET= 0.732717E+07
TAU=  100805.0   TOUT=   730.25   SUMMET= 0.991020E+07
TAU=  104405.7   TOUT=   751.55   SUMMET= 0.125687E+08
TAU=  108006.4   TOUT=   772.95   SUMMET= 0.153052E+08
TAU=  111607.1   TOUT=   794.12   SUMMET= 0.180993E+08
TAU=  114907.7   TOUT=   812.94   SUMMET= 0.206601E+08
************************************************************
DISCHARGING   TAU=   115200.
************************************************************
TAU=  118810.4   TOUT=   879.53   SUMMET= 0.320349E+07
TAU=  122424.1   TOUT=   865.62   SUMMET= 0.635601E+07
TAU=  126037.8   TOUT=   852.14   SUMMET= 0.945903E+07
TAU=  129651.4   TOUT=   839.11   SUMMET= 0.125142E+08
TAU=  133265.1   TOUT=   826.54   SUMMET= 0.155230E+08
TAU=  136878.8   TOUT=   814.44   SUMMET= 0.184828E+08
TAU=  140492.4   TOUT=   802.80   SUMMET= 0.213961E+08
TAU=  144106.1   TOUT=   791.63   SUMMET= 0.242680E+08
TAU=  147719.8   TOUT=   780.93   SUMMET= 0.271011E+08
TAU=  151333.5   TOUT=   770.69   SUMMET= 0.298964E+08
TAU=  154947.1   TOUT=   760.90   SUMMET= 0.326543E+08
TAU=  158560.8   TOUT=   751.57   SUMMET= 0.353790E+08
TAU=  162174.5   TOUT=   742.67   SUMMET= 0.380702E+08
TAU=  165788.1   TOUT=   734.20   SUMMET= 0.407293E+08
TAU=  169401.8   TOUT=   726.15   SUMMET= 0.433587E+08
TAU=  172714.4   TOUT=   719.12   SUMMET= 0.457457E+08
************************************************************
CHARGING   TAU=   172802.
************************************************************
TAU=  176402.8   TOUT=   666.52   SUMMET= 0.236396E+07
TAU=  180003.5   TOUT=   685.18   SUMMET= 0.478821E+07
TAU=  183604.2   TOUT=   705.05   SUMMET= 0.728206E+07
TAU=  187204.9   TOUT=   725.93   SUMMET= 0.984940E+07
TAU=  190805.6   TOUT=   747.42   SUMMET= 0.124931E+08
TAU=  194406.3   TOUT=   769.09   SUMMET= 0.152148E+08
TAU=  198007.0   TOUT=   790.45   SUMMET= 0.180016E+08
TAU=  201307.6   TOUT=   809.56   SUMMET= 0.205624E+08
************************************************************
DISCHARGING   TAU=   201600.
************************************************************
TAU=  205210.3   TOUT=   878.21   SUMMET= 0.319870E+07
TAU=  208824.0   TOUT=   864.29   SUMMET= 0.634642E+07
TAU=  212437.7   TOUT=   850.86   SUMMET= 0.944461E+07
TAU=  216051.4   TOUT=   837.78   SUMMET= 0.124949E+08
TAU=  219665.0   TOUT=   825.22   SUMMET= 0.154990E+08
TAU=  223278.7   TOUT=   813.13   SUMMET= 0.184546E+08
TAU=  226892.4   TOUT=   801.51   SUMMET= 0.213626E+08
TAU=  230506.0   TOUT=   790.37   SUMMET= 0.242306E+08
TAU=  234119.7   TOUT=   779.69   SUMMET= 0.270489E+08
TAU=  237733.4   TOUT=   769.49   SUMMET= 0.298489E+08
TAU=  241347.1   TOUT=   759.74   SUMMET= 0.326031E+08
TAU=  244960.7   TOUT=   750.44   SUMMET= 0.353231E+08
TAU=  248574.4   TOUT=   741.58   SUMMET= 0.380105E+08
TAU=  252188.1   TOUT=   733.15   SUMMET= 0.406670E+08
TAU=  255801.7   TOUT=   725.14   SUMMET= 0.432922E+08
TAU=  259114.3   TOUT=   718.15   SUMMET= 0.456754E+08
```

```
     48.00 TAUH IN H.MIN

    X       TK       TG        UK         UG        ESP        SA
  0.00    795.521  718.948   875072.75  813849.37 0.866419E+11 0.00000
  1.21    776.867  702.142   854553.87  794825.25 0.789050E+11 0.00000
  2.41    757.244  685.918   832968.94  776459.75 0.710001E+11 0.00000
  3.62    737.510  670.558   811261.50  759072.06 0.629076E+11 0.00000
  4.83    718.037  656.259   789840.44  742885.62 0.546117E+11 0.00000
  6.04    699.129  643.183   769041.87  728083.00 0.460989E+11 0.00000
  7.24    681.037  631.462   749140.75  714815.44 0.373579E+11 0.00000
  8.45    663.965  621.207   730361.19  703206.25 0.283802E+11 0.00000
  9.66    648.080  612.507   712888.00  693358.06 0.191607E+11 0.00000
 10.87    633.535  605.435   696888.37  685353.00 0.969838E+10 0.00000
 12.07    620.844  600.000   682928.44  679200.00 0.000000E+00 0.00000
QK-0.109731643E+12
QP,QG,(QG/TAUL)  242.368179        23044064.0        400.140625


**************************************************************
CHARGING    TAU= 172800.
**************************************************************


     49.00 TAUH IN H.MIN

    X       TK       TG        UK         UG        ESP        SA
  0.00    368.767 1000.000   955643.44 1163000.00 0.902186E+11 0.00000
  1.21    830.410  937.542   913451.37 1090361.00 0.797382E+11 0.00000
  2.41    797.990  886.325   877788.56 1030796.56 0.696944E+11 0.00000
  3.62    769.873  843.805   846860.44  981345.75 0.600240E+11 0.00000
  4.83    744.833  807.909   819316.75  939597.87 0.506813E+11 0.00000
  6.04    722.126  776.999   794339.37  903649.31 0.416332E+11 0.00000
  7.24    701.322  749.900   771454.56  872133.62 0.328535E+11 0.00000
  8.45    682.188  725.746   750406.81  844043.25 0.243201E+11 0.00000
  9.66    664.615  704.013   731077.06  818772.50 0.160131E+11 0.00000
 10.87    648.581  684.334   713439.37  795880.62 0.791341E+10 0.00000
 12.07    634.413  666.505   697854.44  775145.50 0.000000E+00 0.00000
QK 0.902186270E+11
QP,QG,(QG/TAUL) -969.635742       -3571206.00       -993.655273


     50.00 TAUH IN H.MIN

    X       TK       TG        UK         UG        ESP        SA
  0.00    915.743 1000.000  1007317.69 1163000.00 0.936135E+11 0.00000
  1.21    872.648  956.065   959913.19 1111904.00 0.825829E+11 0.00000
  2.41    835.184  913.647   918702.75 1062572.00 0.720492E+11 0.00000
  3.62    802.447  874.269   882692.19 1016775.44 0.619484E+11 0.00000
  4.83    773.455  838.427   850800.50  975090.06 0.522283E+11 0.00000
  6.04    747.468  806.050   822214.31  937436.25 0.428474E+11 0.00000
  7.24    723.957  776.845   796353.06  903470.44 0.337717E+11 0.00000
  8.45    702.565  750.443   772822.00  872764.81 0.249730E+11 0.00000
  9.66    683.062  726.500   751367.94  844919.37 0.164266E+11 0.00000
 10.87    665.317  704.805   731849.31  819688.75 0.810987E+10 0.00000
 12.07    649.528  685.105   714480.44  796777.25 0.000000E+00 0.00000
QK 0.936135229E+11
QP,QG,(QG/TAUL) -915.556640       -6956872.00       -973.384033
```

```
      51.00 TAUH IN H.MIN

    X       TK       TG        UK          UG          ESP          SA
  0.00    945.855  1000.000  1040440.25  1163000.00  0.968065E+11  0.00000
  1.21    905.248   969.085   995773.19  1127046.00  0.853891E+11  0.00000
  2.41    867.594   934.760   954352.94  1087126.00  0.744543E+11  0.00000
  3.62    833.290   899.745   916619.06  1046403.37  0.639634E+11  0.00000
  4.83    802.140   865.683   882353.75  1006789.19  0.539762E+11  0.00000
  6.04    773.839   833.419   851222.87   969266.69  0.441557E+11  0.00000
  7.24    748.078   803.337   822886.00   934280.81  0.347686E+11  0.00000
  8.45    724.588   775.505   797046.69   901912.25  0.256853E+11  0.00000
  9.66    703.156   749.915   773472.00   872151.50  0.168791E+11  0.00000
 10.87    683.636   726.449   751999.69   844860.62  0.832544E+10  0.00000
 12.07    666.159   705.039   732775.00   819960.94  0.000000E+00  0.00000
QK 0.968065024E+11
QP,QG,(QG/TAUL) -857.597656        -10140669.0        -948.359863


      52.00 TAUH IN H.MIN

    X       TK       TG        UK          UG          ESP          SA
  0.00    965.172  1000.000  1061689.00  1163000.00  0.997855E+11  0.00000
  1.21    930.027   978.241  1023029.94  1137694.00  0.880961E+11  0.00000
  2.41    894.958   950.939   984453.81  1105942.00  0.768398E+11  0.00000
  3.62    861.319   920.708   947450.62  1070783.00  0.660072E+11  0.00000
  4.83    829.645   889.473   912609.31  1034457.44  0.555775E+11  0.00000
  6.04    800.142   858.499   880156.00   998434.31  0.455250E+11  0.00000
  7.24    772.833   828.604   850116.87   963666.69  0.358230E+11  0.00000
  8.45    747.655   800.205   822420.50   930638.94  0.264448E+11  0.00000
  9.66    724.509   773.597   796960.44   899693.25  0.173466E+11  0.00000
 10.87    703.305   748.792   773635.44   870845.00  0.855790E+10  0.00000
 12.07    684.178   725.941   752595.81   844269.69  0.000000E+00  0.00000
QK 0.997855068E+11
QP,QG,(QG/TAUL) -796.825683        -13110546.0        -920.761718


      53.00 TAUH IN H.MIN

    X       TK       TG        UK          UG          ESP          SA
  0.00    977.574  1000.000  1075332.00  1163000.00  0.102542E+12  0.00000
  1.21    948.638   984.679  1043501.81  1145182.00  0.906616E+11  0.00000
  2.41    917.528   963.254  1009281.06  1120265.00  0.791512E+11  0.00000
  3.62    886.029   937.727   974631.87  1090576.00  0.680271E+11  0.00000
  4.83    855.151   909.846   940665.81  1058151.00  0.572876E+11  0.00000
  6.04    825.508   880.981   908059.37  1024581.00  0.469214E+11  0.00000
  7.24    797.448   852.115   877192.37   991010.19  0.369111E+11  0.00000
  8.45    771.139   823.963   848252.94   958268.75  0.272362E+11  0.00000
  9.66    746.649   796.931   821313.87   926831.31  0.178746E+11  0.00000
 10.87    723.990   771.358   796389.25   897089.81  0.880378E+10  0.00000
 12.07    703.356   747.383   773692.06   869207.00  0.000000E+00  0.00000
QK 0.102542278E+12
QP,QG,(QG/TAUL) -734.482666        -15858446.0        -891.687500
```

```
     54.00 TAUH IN H.MIN

    X        TK        TG         UK          UG          ESP         SA
  0.00     985.543   1000.000  1084098.00  1163000.00  0.105073E+12  0.00000
  1.21     962.485    989.209  1058734.00  1150450.00  0.930577E+11  0.00000
  2.41     935.811    972.573  1029392.12  1131103.00  0.813492E+11  0.00000
  3.62     907.309    951.388   998040.12  1106465.00  0.699810E+11  0.00000
  4.83     878.189    927.040   966008.06  1078147.00  0.589682E+11  0.00000
  6.04     849.315    900.760   934246.44  1047584.37  0.483130E+11  0.00000
  7.24     821.275    873.578   903403.06  1015971.37  0.380090E+11  0.00000
  8.45     794.453    846.284   873898.19   984228.87  0.280432E+11  0.00000
  9.66     769.082    819.503   845990.62   953082.62  0.183995E+11  0.00000
 10.87     745.303    793.609   819833.31   922967.75  0.905885E+10  0.00000
 12.07     723.398    769.046   795738.50   894400.31  0.000000E+00  0.00000
OK 0.105073016E+12
OP,QG,(QG/TAUL) -671.499267      -18282384.0       -857.100341


     55.00 TAUH IN H.MIN

    X        TK        TG         UK          UG          ESP         SA
  0.00     990.669   1000.000  1089736.00  1163000.00  0.107378E+12  0.00000
  1.21     972.713    992.396  1058734.00  1154157.00  0.952684E+11  0.00000
  2.41     950.412    979.592  1045453.00  1139266.00  0.834067E+11  0.00000
  3.62     925.299    962.260  1017828.69  1119108.00  0.718375E+11  0.00000
  4.83     898.563    941.360   988420.06  1094802.00  0.605881E+11  0.00000
  6.04     871.162    917.900   958278.62  1067518.00  0.496725E+11  0.00000
  7.24     843.827    892.815   928209.37  1038343.56  0.390946E+11  0.00000
  8.45     817.094    866.923   898803.44  1008231.44  0.288501E+11  0.00000
  9.66     791.345    840.885   870479.31   977949.06  0.189294E+11  0.00000
 10.87     766.842    815.235   843526.62   948118.19  0.931861E+10  0.00000
 12.07     743.974    790.381   818371.31   919213.00  0.000000E+00  0.00000
OK 0.107378376E+12
OP,QG,(QG/TAUL) -609.467285      -20465984.0       -822.704834


     56.00 TAUH IN H.MIN

    X        TK        TG         UK          UG          ESP         SA
  0.00     993.970   1000.000  1093367.00  1163000.00  0.109463E+12  0.00000
  1.21     980.220    994.640  1078242.00  1156767.00  0.972859E+11  0.00000
  2.41     961.935    984.855  1058129.00  1145387.00  0.853063E+11  0.00000
  3.62     940.274    970.840  1034301.50  1129087.00  0.735742E+11  0.00000
  4.83     916.266    953.163  1007892.44  1108529.00  0.621232E+11  0.00000
  6.04     890.833    932.556   979916.62  1084563.00  0.509771E+11  0.00000
  7.24     864.754    909.827   951229.75  1058129.00  0.401438E+11  0.00000
  8.45     838.655    885.669   922520.50  1030033.25  0.296423E+11  0.00000
  9.66     813.021    860.856   894322.56  1001175.37  0.194548E+11  0.00000
 10.87     788.218    835.812   867040.12   972049.75  0.957851E+10  0.00000
 12.07     764.736    811.200   841209.37   943426.25  0.000000E+00  0.00000
OK 0.109462552E+12
OP,QG,(QG/TAUL) -548.934326      -22421056.0       -788.856933
```

```
*************************************************************
DISCHARGING   TAU=   201600.
*************************************************************


    57.00 TAUH IN H.MIN

     X        TK        TG        UK          UG          ESP          SA
   0.00    980.450   878.252  1078495.00   994181.25  0.107373E+12  0.00000
   1.21    965.520   854.572  1062072.00   967375.37  0.980376E+11  0.00000
   2.41    946.365   829.233  1041001.31   938692.12  0.883923E+11  0.00000
   3.62    924.092   802.740  1016501.50   908701.50  0.784271E+11  0.00000
   4.83    899.650   775.462   989615.62   877823.56  0.681366E+11  0.00000
   6.04    873.875   747.645   961262.56   846333.87  0.575196E+11  0.00000
   7.24    847.457   719.397   932202.75   814358.06  0.465806E+11  0.00000
   8.45    820.937   690.707   903030.81   781880.87  0.353318E+11  0.00000
   9.66    794.715   661.439   874186.31   748749.31  0.237950E+11  0.00000
  10.87    769.069   631.340   845976.44   714677.19  0.120026E+11  0.00000
  12.07    744.463   600.000   818915.00   679200.00  0.000000E+00  0.00000
QK-0.109463011E+12
QP,QG,(QG/TAUL)   566.965332        2094467.00       581.781005


    58.00 TAUH IN H.MIN

     X        TK        TG        UK          UG          ESP          SA
   0.00    966.869   864.376  1063556.00   978473.37  0.105383E+12  0.00000
   1.21    950.901   840.752  1045990.94   951731.94  0.962642E+11  0.00000
   2.41    931.002   815.707  1024102.12   923380.44  0.868296E+11  0.00000
   3.62    908.242   789.733   999066.69   893978.19  0.770712E+11  0.00000
   4.83    883.499   763.205   971849.37   863948.37  0.669850E+11  0.00000
   6.04    857.536   736.384   943289.69   833586.87  0.565704E+11  0.00000
   7.24    830.975   709.425   914072.25   803068.81  0.458319E+11  0.00000
   8.45    804.292   682.382   884721.19   772456.06  0.347805E+11  0.00000
   9.66    777.828   655.223   855610.94   741712.37  0.234361E+11  0.00000
  10.87    751.807   627.837   826987.87   710711.62  0.118287E+11  0.00000
  12.07    726.684   600.000   799352.75   679200.00  0.000000E+00  0.00000
QK-0.109463011E+12
QP,QG,(QG/TAUL)   538.691162        4083876.00       567.243408


    59.00 TAUH IN H.MIN

     X        TK        TG        UK          UG          ESP          SA
   0.00    953.276   850.941  1048604.00   963264.87  0.103494E+12  0.00000
   1.21    936.415   827.468  1030056.69   936693.44  0.945668E+11  0.00000
   2.41    915.906   802.799  1007496.87   908769.06  0.853237E+11  0.00000
   3.62    892.789   777.417   982067.75   880036.69  0.757571E+11  0.00000
   4.83    867.878   751.695   954665.75   850918.75  0.658633E+11  0.00000
   6.04    841.875   725.906   926063.06   821726.12  0.556417E+11  0.00000
   7.24    815.348   700.238   896882.87   792669.56  0.450961E+11  0.00000
   8.45    788.723   674.798   867595.87   763870.94  0.342364E+11  0.00000
   9.66    762.301   649.629   838531.37   735380.37  0.230305E+11  0.00000
  10.87    736.272   624.727   809899.25   707190.62  0.116555E+11  0.00000
  12.07    711.102   600.000   782211.87   679200.00  0.000000E+00  0.00000
QK-0.109463011E+12
QP,QG,(QG/TAUL)   511.316162        5973109.00       553.125732
```

```
      60.00 TAUH IN H.MIN

    X        TK        TG         UK          UG         ESP          SA
  0.00    939.729   837.957   1033701.50   948567.69   0.101702E+12   0.00000
  1.21    922.113   814.721   1014324.31   922264.37   0.929447E+11   0.00000
  2.41    901.121   790.502    991232.94   894848.19   0.838758E+11   0.00000
  3.62    877.767   765.772    965543.44   866853.87   0.744868E+11   0.00000
  4.83    852.811   740.899    938092.12   838697.81   0.647739E+11   0.00000
  6.04    826.903   716.167    909593.87   810700.50   0.547358E+11   0.00000
  7.24    800.565   691.782    880621.50   783097.37   0.443754E+11   0.00000
  8.45    774.186   667.892    851605.06   756054.25   0.337014E+11   0.00000
  9.66    748.039   644.597    822843.25   729683.75   0.227293E+11   0.00000
 10.87    722.297   621.965    794526.31   704064.19   0.114837E+11   0.00000
 12.07    697.446   600.000    767191.12   679200.00   0.000000E+00   0.00000
QK-0.109463011E+12
QP,QG,(QG/TAUL)   484.861084         7765432.00         539.335693


      61.00 TAUH IN H.MIN

    X        TK        TG         UK          UG         ESP          SA
  0.00    926.277   825.437   1018904.69   934394.87   0.100003E+12   0.00000
  1.21    908.038   802.511    998841.94   908442.31   0.913971E+11   0.00000
  2.41    886.682   778.804    975350.31   881606.12   0.824864E+11   0.00000
  3.62    863.204   754.775    949524.62   854404.94   0.732618E+11   0.00000
  4.83    838.316   730.784    922148.06   827248.00   0.637185E+11   0.00000
  6.04    812.623   707.120    893885.75   800459.94   0.538546E+11   0.00000
  7.24    786.608   684.004    865269.12   774292.19   0.436717E+11   0.00000
  8.45    760.633   661.607    836696.75   748939.56   0.331768E+11   0.00000
  9.66    734.952   640.070    808447.06   724559.25   0.223836E+11   0.00000
 10.87    709.729   619.512    780701.56   701287.94   0.113139E+11   0.00000
 12.07    685.479   600.000    754027.00   679200.00   0.000000E+00   0.00000
QK-0.109463011E+12
QP,QG,(QG/TAUL)   459.349853         9464217.00         525.864257


      62.00 TAUH IN H.MIN

    X        TK        TG         UK          UG         ESP          SA
  0.00    912.968   813.383   1004264.75   920750.00   0.983937E+11   0.00000
  1.21    894.229   790.834    983651.50   895224.69   0.899226E+11   0.00000
  2.41    872.620   767.692    959882.44   869027.44   0.811557E+11   0.00000
  3.62    849.123   744.403    934035.50   842663.87   0.720828E+11   0.00000
  4.83    824.406   721.318    906846.12   816532.62   0.626985E+11   0.00000
  6.04    799.033   698.725    878936.12   790956.87   0.529995E+11   0.00000
  7.24    773.456   676.852    850302.00   766197.06   0.429863E+11   0.00000
  8.45    748.016   655.889    822817.87   742466.75   0.326641E+11   0.00000
  9.66    722.953   635.999    795248.37   719950.69   0.220445E+11   0.00000
 10.87    698.430   617.334    768273.37   698822.62   0.111467E+11   0.00000
 12.07    674.990   600.000    742488.75   679200.00   0.000000E+00   0.00000
QK-0.109463011E+12
QP,QG,(QG/TAUL)   434.789306        11072866.0         512.709472
```

```
    63.00 TAUH IN H.MIN

   X       TK       TG        UK          UG         ESP          SA
 0.00    899.842   801.800   989826.56   907637.69  0.968716E+11  0.00000
 1.21    880.717   779.684   968789.00   882602.81  0.895199E+11  0.00000
 2.41    858.961   757.152   944856.81   857095.75  0.788833E+11  0.00000
 3.62    835.540   734.632   919093.87   831603.69  0.709506E+11  0.00000
 4.83    811.084   712.469   892192.81   806514.37  0.617150E+11  0.00000
 6.04    786.124   690.941   864736.87   782145.25  0.521719E+11  0.00000
 7.24    761.084   670.282   837192.81   758759.44  0.423204E+11  0.00000
 8.45    736.286   650.689   809914.75   736579.69  0.321642E+11  0.00000
 9.66    711.962   632.337   783157.94   715806.12  0.217126E+11  0.00000
10.87    688.276   615.400   757104.19   696633.31  0.109824E+11  0.00000
12.07    665.795   600.000   732374.44   679200.00  0.000000E+00  0.00000
QK-0.109463011E+12
QP,QG,(QG/TAUL)   411.187255        12594820.0        499.872314


    64.00 TAUH IN H.MIN

   X       TK       TG        UK          UG         ESP          SA
 0.00    886.936   790.686   975629.50   895056.12  0.954327E+11  0.00000
 1.21    867.531   769.054   954284.87   870569.00  0.871869E+11  0.00000
 2.41    845.723   747.165   930295.25   845791.37  0.786687E+11  0.00000
 3.62    822.466   725.439   904712.31   821196.69  0.698652E+11  0.00000
 4.83    798.354   704.202   878189.50   797156.87  0.607685E+11  0.00000
 6.04    773.887   683.729   851275.69   773981.19  0.513725E+11  0.00000
 7.24    749.465   664.249   824411.37   751939.75  0.416751E+11  0.00000
 8.45    725.394   645.960   797933.94   731227.31  0.316780E+11  0.00000
 9.66    701.901   629.046   772091.62   712080.00  0.213887E+11  0.00000
10.87    679.154   613.683   747069.44   694689.06  0.108214E+11  0.00000
12.07    657.734   600.000   723507.56   679200.00  0.000000E+00  0.00000
QK-0.109463011E+12
QP,QG,(QG/TAUL)   388.540283        14033529.0        487.353271


    65.00 TAUH IN H.MIN

   X       TK       TG        UK          UG         ESP          SA
 0.00    874.280   780.039   961707.87   883003.69  0.940737E+11  0.00000
 1.21    854.695   758.931   940164.75   859109.62  0.859220E+11  0.00000
 2.41    832.922   737.717   916214.62   835095.25  0.775111E+11  0.00000
 3.62    809.907   716.798   890898.19   811415.06  0.688267E+11  0.00000
 4.83    786.212   696.488   864833.00   788424.56  0.598594E+11  0.00000
 6.04    762.306   677.052   838536.50   766422.50  0.506021E+11  0.00000
 7.24    738.569   658.712   812425.56   745661.62  0.410509E+11  0.00000
 8.45    715.293   641.663   786822.75   726362.94  0.312062E+11  0.00000
 9.66    692.700   626.086   761970.37   708730.00  0.210733E+11  0.00000
10.87    670.960   612.157   738056.62   692962.19  0.106642E+11  0.00000
12.07    650.667   600.000   715733.31   679200.00  0.000000E+00  0.00000
QK-0.109463011E+12
QP,QG,(QG/TAUL)   366.845947        15392423.0        475.152343
```

```
66.00 TAUH IN H.MIN

   X       TK       TG       UK         UG         ESP           SA
 0.00   861.901  769.852  948091.25  871472.94  0.927910E+11  0.00000
 1.21   842.227  749.305  926449.44  848213.50  0.847230E+11  0.00000
 2.41   820.570  728.786  902627.12  824985.94  0.764093E+11  0.00000
 3.62   797.868  708.684  877654.75  802230.31  0.678345E+11  0.00000
 4.83   774.651  689.296  852116.25  780283.00  0.589878E+11  0.00000
 6.04   751.364  670.874  826500.25  759429.19  0.498609E+11  0.00000
 7.24   728.365  653.632  801201.75  739912.06  0.404486E+11  0.00000
 8.45   705.936  637.759  776529.56  721943.12  0.307494E+11  0.00000
 9.66   684.291  623.426  752720.19  705718.81  0.207670E+11  0.00000
10.87   663.603  610.802  729963.75  691428.50  0.105109E+11  0.00000
12.07   644.469  600.000  708916.37  679200.00  0.000000E+00  0.00000
QK-0.109463011E+12
QP,QG,(QG/TAUL)  346.090576        16674913.0        463.269043


67.00 TAUH IN H.MIN

   X       TK       TG       UK         UG         ESP           SA
 0.00   849.822  760.122  934804.56  860458.06  0.915814E+11  0.00000
 1.21   830.141  740.162  913155.44  837863.81  0.835876E+11  0.00000
 2.41   808.674  720.356  889541.06  815443.25  0.753620E+11  0.00000
 3.62   786.347  701.073  864981.44  793614.44  0.668880E+11  0.00000
 4.83   763.663  682.595  840029.12  772698.31  0.581536E+11  0.00000
 6.04   741.041  665.162  815145.56  752963.50  0.491492E+11  0.00000
 7.24   718.823  648.976  790705.44  734640.44  0.398683E+11  0.00000
 8.45   697.276  634.213  767004.12  717928.94  0.303080E+11  0.00000
 9.66   676.611  621.036  744271.87  703012.62  0.204700E+11  0.00000
10.87   656.999  609.599  722698.94  690066.31  0.103619E+11  0.00000
12.07   639.035  600.000  702938.19  679200.00  0.000000E+00  0.00000
QK-0.109463011E+12
QP,QG,(QG/TAUL)  326.263671        17877312.0        451.523193


68.00 TAUH IN H.MIN

   X       TK       TG       UK         UG         ESP           SA
 0.00   838.062  750.839  921868.75  849949.62  0.904416E+11  0.00000
 1.21   818.450  731.489  900295.50  828046.00  0.825137E+11  0.00000
 2.41   797.237  712.406  876961.00  806443.25  0.743677E+11  0.00000
 3.62   775.340  693.940  852824.37  785539.94  0.652865E+11  0.00000
 4.83   753.235  676.359  828558.56  765638.25  0.573964E+11  0.00000
 6.04   731.317  659.885  804449.37  746989.69  0.484670E+11  0.00000
 7.24   709.911  644.708  780901.75  729809.44  0.393104E+11  0.00000
 8.45   689.271  630.993  758198.00  714284.06  0.298823E+11  0.00000
 9.66   669.601  618.887  736560.87  700580.50  0.201837E+11  0.00000
10.87   651.072  608.530  716179.06  688856.31  0.102173E+11  0.00000
12.07   634.268  600.000  697695.00  679200.00  0.000000E+00  0.00000
QK-0.109463011E+12
QP,QG,(QG/TAUL)  307.348877        19009056.0        440.099365
```

```
    69.00 TAUH IN H.MIN

    X        TK       TG        UK          UG          ESP          SA
  0.00    824.637  741.292   903300.87   839935.44  0.893682E+11  0.00000
  1.21    809.182  723.271   887878.75   818743.06  0.814989E+11  0.00000
  2.41    786.262  704.915   864888.25   797964.25  0.734254E+11  0.00000
  3.62    764.843  687.261   841326.87   777980.19  0.651290E+11  0.00000
  4.83    743.354  670.558   817689.44   759071.25  0.565958E+11  0.00000
  6.04    722.170  655.012   794387.12   741473.44  0.478141E+11  0.00000
  7.24    701.596  640.799   771755.87   725384.62  0.387749E+11  0.00000
  8.45    681.877  628.070   750064.44   710975.19  0.294724E+11  0.00000
  9.66    663.207  616.957   729527.31   698395.19  0.199053E+11  0.00000
 10.87    645.754  607.581   710329.06   687781.37  0.100772E+11  0.00000
 12.07    630.087  600.000   693095.94   679200.00  0.000000E+00  0.00000
QK-0.109463011E+12
QP,QG,(QG/TAUL)  289.323242      20073984.0       429.004882


    70.00 TAUH IN H.MIN

    X        TK       TG        UK          UG          ESP          SA
  0.00    815.558  733.573   897114.44   830405.25  0.883581E+11  0.00000
  1.21    796.283  715.492   875911.06   809936.69  0.805409E+11  0.00000
  2.41    775.746  697.866   853321.19   789984.25  0.725327E+11  0.00000
  3.62    754.845  681.013   830329.81   770907.19  0.643144E+11  0.00000
  4.83    734.004  665.166   807404.69   752967.50  0.558711E+11  0.00000
  6.04    713.576  650.515   784933.37   736382.81  0.471902E+11  0.00000
  7.24    693.848  637.220   763233.12   721333.37  0.382616E+11  0.00000
  8.45    675.053  625.417   742558.75   707972.62  0.290785E+11  0.00000
  9.66    657.377  615.222   723115.31   696431.87  0.196379E+11  0.00000
 10.87    640.983  606.737   705081.37   686826.62  0.994173E+10  0.00000
 12.07    626.419  600.000   689061.37   679200.00  0.000000E+00  0.00000
QK-0.109463011E+12
QP,QG,(QG/TAUL)  272.168945      21075680.0       418.240722


    71.00 TAUH IN H.MIN

    X        TK       TG        UK          UG          ESP          SA
  0.00    804.836  725.570   885319.87   821344.87  0.874082E+11  0.00000
  1.21    785.814  708.136   864395.62   801609.75  0.796372E+11  0.00000
  2.41    765.687  691.237   842255.87   782480.62  0.716882E+11  0.00000
  3.62    745.338  675.173   819872.06   764295.75  0.635415E+11  0.00000
  4.83    725.169  660.158   797685.94   747298.56  0.551815E+11  0.00000
  6.04    705.511  646.367   776062.19   731687.62  0.465948E+11  0.00000
  7.24    686.636  633.945   755299.44   717625.87  0.377705E+11  0.00000
  8.45    668.762  623.011   735638.31   705248.50  0.287005E+11  0.00000
  9.66    652.066  613.664   717273.31   694668.12  0.193806E+11  0.00000
 10.87    636.704  605.988   700375.00   685978.37  0.981103E+10  0.00000
 12.07    623.201  600.000   685521.69   679200.00  0.000000E+00  0.00000
QK-0.109463011E+12
QP,QG,(QG/TAUL)  255.860305      22017760.0       407.807373


    72.00 TAUH IN H.MIN

    X        TK       TG        UK          UG          ESP          SA
  0.00    794.486  717.976   873935.06   812748.75  0.865164E+11  0.00000
  1.21    775.766  701.194   853343.00   793751.31  0.787865E+11  0.00000
  2.41    756.087  685.015   831695.75   775437.00  0.708909E+11  0.00000
  3.62    736.318  669.723   809949.87   758126.19  0.628097E+11  0.00000
  4.83    716.838  655.514   788522.06   742041.62  0.545267E+11  0.00000
  6.04    697.959  642.547   767754.75   727363.19  0.460280E+11  0.00000
  7.24    679.935  630.951   747928.12   714237.25  0.373016E+11  0.00000
  8.45    662.971  620.830   729268.37   702780.00  0.283386E+11  0.00000
  9.66    647.235  612.266   711958.19   693085.25  0.191336E+11  0.00000
 10.87    632.871  605.323   696158.62   685225.50  0.968521E+10  0.00000
 12.07    620.380  600.000   682418.56   679200.00  0.000000E+00  0.00000
QK-0.109463011E+12
QP,QG,(QG/TAUL)  240.387237      22901328.0       397.662109
```

NUMERICAL OUTPUT No. V

```
        11          =M NUMBER OF CALCULATION POINTS IN X-DIRECTION
 )00.00000          =TGEBE [C] INLET TEMPERATURE FLUID (CHARGING)
 )0.000000          =TGEEN [C] INLET TEMPERATURE FLUID (DISCHARGING)
 30.000000          =TKS [C] MELTING POINT
 30.000000          =TKK [C] CRYSTALLIZATION POINT
 ).0000000          =TU [C] AMBIENT TEMPERATUR
 500.00000          =RHOK [KG/M**3] DENSITY SOLID MEDIUM
 l00.00000          =CKK [J/(KG*K)] SPECIFIC HEAT CERAMIC
 433.33007          =CKS [J/(KG*K)] SPECIFIC HEAT PHASE-CHANGE-MATERIAL
 .00000000          =RLAMK [W/(M*K)] COEFF. OF THERMAL CONDUCT. SOLID MED.
 20000.000          =RK [J/KG] HEAT OF FUSION
 449999988          =RMS MASS-SHARE OF SALT
 310500025          =RHOGB [KG/M**3] DENSITY FLUID (CHARGING)
 .66800022          =RHOGE [KG/M**3] DENSITY FLUID (DISCHARGING)
 l63.00000          =CGB [J/(KG*K)] SPECIFIC HEAT FLUID (CHARGING)
 132.00000          =CGE [J/(KG*K)] SPECIFIC HEAT FLUID (DISCHARGING)
 727999806E-01      =RLAMGB [W/(M*K)] COEFF. OF THERMAL COND. FLUID CHARG.
 665000081E-01      =RLAMGE [W/(M*K)] COEFF. OF THERM. COND. FLUID DISCH.
 709999978          =PRGB PRANDTL-NUMBER (CHARGING)
 9.0000000          =PRGE PRANDTL-NUMBER (DISCHARGING)
 00.000000          =TKVA NOMINAL TEMPERATURE FOR HEAT LOSS
 .00000000          =DTKVU TEMPERATURE GRADIENT FOR HEAT LOSS
 00000.000          =QPDIS[W] ENERGY OUTPUT (DISCHARGING)
 7600.0000          =TAUDIS [S] DISCHARGING-TIME
 8800.0000          =TAUCHA [S] CHARGING-TIME
 699999983          =ETASTO EFFICIENCY OF STORING (STARTVALUE)
 00.000000          =TCOUT [C] OUTLET-TEMPERATURE FLUID (CHARGING)
 20.000000          =TDOUT [C] OUTLET-TEMPERATURE FLUID (DISCHARGING)
 700000000          =FF FILLING FACTOR
 143700002E-03      =RNUEB [M**2/S] VISCOSITY FLUID (CHARGING)
 110500003E-02      =RNUEE [M**2/S] VISCOSITY FLUID (DISCHARGING)
 00.000000          =DELTAT [C] AVERAGE TEMPERATURE DIFFERENCE CERAMIC
```

```
=   0.416178107
=   0.100180625
    0.199999986E-02
    0.118000009E-07
=   7808.19140
=   159.106323
=   0.234460413
=   0.181297697E-01
T   0.246857113E+11
ST    46664.8984
      18.6659545
CHA   857143.000
PGB   2.50000000
PGE   1.19999980
      9.46819782
      1.89363956
RIT  0.349841937E-01
     0.349999964E-01
T     2321.04687
L     12.2645540
UT    10.5602321
UL     2.70995616
PHAT   21.9652710
PHAL    5.14891624
PHAB   21.2138824
PHAE    5.10651874
OHR       879
   0.845696687
    1.97064113
   96.6510467
   0.483749248E-01
```

```
TAU=    3601.7   TOUT=   694.00   SUMMET= 0.239725E+07
TAU=    7203.3   TOUT=   742.41   SUMMET= 0.496739E+07
TAU=   10805.8   TOUT=   786.05   SUMMET= 0.770367E+07
TAU=   14407.7   TOUT=   822.15   SUMMET= 0.105846E+08
TAU=   18309.5   TOUT=   851.14   SUMMET= 0.138337E+08
TAU=   21611.0   TOUT=   867.35   SUMMET= 0.166556E+08
TAU=   25212.7   TOUT=   877.63   SUMMET= 0.195613E+08
TAU=   28514.2   TOUT=   882.15   SUMMET= 0.222172E+08
*************************************************************
DISCHARGING   TAU=  28300.
*************************************************************
TAU=   32410.2   TOUT=   944.26   SUMMET= 0.344783E+07
TAU=   36323.8   TOUT=   921.13   SUMMET= 0.709738E+07
TAU=   39635.3   TOUT=   901.47   SUMMET= 0.101144E+08
TAU=   43247.9   TOUT=   880.60   SUMMET= 0.133325E+08
TAU=   46860.4   TOUT=   860.89   SUMMET= 0.164768E+08
TAU=   50473.0   TOUT=   842.49   SUMMET= 0.195434E+08
TAU=   54085.6   TOUT=   825.50   SUMMET= 0.225433E+08
TAU=   57698.2   TOUT=   810.15   SUMMET= 0.254862E+08
TAU=   61310.7   TOUT=   797.42   SUMMET= 0.283779E+08
TAU=   64923.3   TOUT=   785.45   SUMMET= 0.312242E+08
TAU=   68535.9   TOUT=   775.73   SUMMET= 0.340312E+08
TAU=   72148.4   TOUT=   768.28   SUMMET= 0.368062E+08
TAU=   75761.0   TOUT=   762.83   SUMMET= 0.395600E+08
TAU=   79373.6   TOUT=   759.69   SUMMET= 0.422995E+08
TAU=   82986.2   TOUT=   756.47   SUMMET= 0.450283E+08
TAU=   86297.7   TOUT=   753.88   SUMMET= 0.475178E+08
*************************************************************
CHARGING     TAU=  86400.
*************************************************************
TAU=   90002.0   TOUT=   664.75   SUMMET= 0.229877E+07
TAU=   93603.7   TOUT=   716.64   SUMMET= 0.476797E+07
TAU=   97205.4   TOUT=   767.04   SUMMET= 0.742326E+07
TAU=  100807.1   TOUT=   809.85   SUMMET= 0.102480E+08
TAU=  104408.7   TOUT=   842.20   SUMMET= 0.132082E+08
TAU=  108010.4   TOUT=   863.49   SUMMET= 0.162638E+08
TAU=  111612.1   TOUT=   875.96   SUMMET= 0.191943E+08
TAU=  114913.6   TOUT=   881.70   SUMMET= 0.218502E+08
*************************************************************
DISCHARGING   TAU=  115200.
*************************************************************
TAU=  118810.1   TOUT=   944.40   SUMMET= 0.344839E+07
TAU=  122422.7   TOUT=   923.03   SUMMET= 0.682089E+07
TAU=  126035.3   TOUT=   901.52   SUMMET= 0.101157E+08
TAU=  129647.8   TOUT=   880.59   SUMMET= 0.133338E+08
TAU=  133260.4   TOUT=   860.85   SUMMET= 0.164780E+08
TAU=  136873.0   TOUT=   842.38   SUMMET= 0.195443E+08
TAU=  140485.6   TOUT=   825.41   SUMMET= 0.225440E+08
TAU=  144098.1   TOUT=   810.04   SUMMET= 0.254864E+08
TAU=  147710.7   TOUT=   797.33   SUMMET= 0.283777E+08
TAU=  151323.3   TOUT=   785.40   SUMMET= 0.312239E+08
TAU=  154935.8   TOUT=   775.73   SUMMET= 0.340307E+08
TAU=  158548.4   TOUT=   768.25   SUMMET= 0.368055E+08
TAU=  162161.0   TOUT=   762.80   SUMMET= 0.395592E+08
TAU=  165773.6   TOUT=   759.67   SUMMET= 0.422987E+08
TAU=  169386.1   TOUT=   756.40   SUMMET= 0.450271E+08
TAU=  172697.6   TOUT=   753.83   SUMMET= 0.475166E+08
*************************************************************
CHARGING     TAU=  172800.
*************************************************************
TAU=  176402.0   TOUT=   664.63   SUMMET= 0.229835E+07
TAU=  180003.7   TOUT=   716.53   SUMMET= 0.476713E+07
TAU=  183605.5   TOUT=   766.96   SUMMET= 0.742202E+07
TAU=  187207.1   TOUT=   809.70   SUMMET= 0.102465E+08
TAU=  190808.8   TOUT=   842.09   SUMMET= 0.132064E+08
TAU=  194410.5   TOUT=   863.43   SUMMET= 0.162619E+08
TAU=  198012.1   TOUT=   875.88   SUMMET= 0.191924E+08
TAU=  201313.7   TOUT=   881.69   SUMMET= 0.218483E+08
*************************************************************
DISCHARGING   TAU=  201600.
*************************************************************
TAU=  205210.2   TOUT=   944.37   SUMMET= 0.344829E+07
TAU=  208822.8   TOUT=   923.00   SUMMET= 0.682067E+07
TAU=  212435.3   TOUT=   901.49   SUMMET= 0.101154E+08
TAU=  216047.9   TOUT=   880.55   SUMMET= 0.133334E+08
TAU=  219660.5   TOUT=   860.82   SUMMET= 0.164774E+08
TAU=  223273.1   TOUT=   842.35   SUMMET= 0.195432E+08
TAU=  226885.6   TOUT=   825.37   SUMMET= 0.225432E+08
TAU=  230498.2   TOUT=   810.01   SUMMET= 0.254854E+08
TAU=  234110.8   TOUT=   797.30   SUMMET= 0.283766E+08
TAU=  237723.3   TOUT=   785.37   SUMMET= 0.312228E+08
TAU=  241335.9   TOUT=   775.70   SUMMET= 0.340294E+08
TAU=  244948.5   TOUT=   768.23   SUMMET= 0.368041E+08
TAU=  248561.1   TOUT=   762.78   SUMMET= 0.395577E+08
TAU=  252173.6   TOUT=   759.66   SUMMET= 0.422972E+08
TAU=  255786.2   TOUT=   756.39   SUMMET= 0.450255E+08
TAU=  257592.5   TOUT=   754.85   SUMMET= 0.463835E+08
```

```
        48.00 TAUH IN H.MIN

   X      TK       TG       UK          UG        ESP         SA
 0.00   780.000  753.766  1246869.00   853263.00  0.448718E+11  0.97445
 0.95   780.000  742.867  1222774.00   840925.81  0.413405E+11  0.88808
 1.89   780.000  727.437  1184165.00   823458.50  0.377702E+11  0.74970
 2.84   780.000  705.596  1125682.00   798735.31  0.341404E+11  0.54008
 3.79   780.000  674.680  1037486.25   763737.81  0.304021E+11  0.22397
 4.73   732.332  640.814   915413.25   725401.25  0.263616E+11  0.00000
 5.68   653.601  619.138   816999.69   700864.69  0.218068E+11  0.00000
 6.63   628.685  609.992   785854.94   690511.25  0.167616E+11  0.00000
 7.57   616.344  604.787   770429.44   684619.50  0.113741E+11  0.00000
 8.52   608.307  601.654   760383.00   681072.87  0.576020E+10  0.00000
 9.47   602.964  600.000   753704.06   679200.00  0.000000E+00  0.00000
QK-0.627463454E+11
QP,QG,(QG/TAUL)  208.875305        17852576.0         310.200683


********************************************************
CHARGING    TAU=  172800.
********************************************************


        49.00 TAUH IN H.MIN

   X      TK       TG       UK          UG         ESP         SA
 0.00   930.247 1000.000  1441807.00  1163000.00  0.486266E+11  1.00000
 0.95   880.000  952.506  1356481.00  1107765.00  0.420998E+11  0.91929
 1.89   852.640  909.362  1274966.00  1057588.00  0.359623E+11  0.74970
 2.84   828.602  874.952  1186434.00  1017569.12  0.302214E+11  0.54008
 3.79   811.875  847.562  1077330.00   985715.12  0.249415E+11  0.22397
 4.73   781.233  822.028   976539.94   956018.87  0.201512E+11  0.00000
 5.68   733.196  789.580   916493.75   918282.19  0.157361E+11  0.00000
 6.63   697.422  752.403   871776.06   875044.31  0.115653E+11  0.00000
 7.57   669.812  717.974   837263.81   835003.37  0.757935E+10  0.00000
 8.52   648.617  688.542   810770.19   800749.37  0.373564E+10  0.00000
 9.47   632.744  664.612   790929.12   772943.62  0.000000E+00  0.00000
QK 0.486265610E+11
QP,QG,(QG/TAUL) -975.141113       -3741531.00        -1040.54858


        50.00 TAUH IN H.MIN

   X      TK       TG       UK          UG         ESP         SA
 0.00   978.764 1000.000  1502453.00  1163000.00  0.518685E+11  1.00000
 0.95   935.792  978.613  1448739.00  1138127.00  0.449852E+11  1.00000
 1.89   880.000  943.212  1355236.00  1096956.00  0.384452E+11  0.91483
 2.84   878.023  911.074  1248212.00  1059579.00  0.323730E+11  0.54008
 3.79   855.263  888.837  1131565.00  1033717.50  0.268235E+11  0.22397
 4.73   828.486  865.322  1035606.12  1006369.50  0.217680E+11  0.00000
 5.68   793.474  838.121   991841.19   974735.12  0.170394E+11  0.00000
 6.63   759.345  807.231   949179.75   938810.25  0.125124E+11  0.00000
 7.57   727.869  775.379   909835.75   901766.00  0.817658E+10  0.00000
 8.52   700.098  744.643   875120.87   866019.56  0.401352E+10  0.00000
 9.47   676.579  716.456   845723.06   833238.75  0.000000E+00  0.00000
QK 0.518685450E+11
QP,QG,(QG/TAUL) -824.403076       -6964807.00        -975.546386
```

```
      51.00 TAUH IN H.MIN
 183599.088738381862
   X       TK        TG        UK          UG         ESP         SA
 0.00    993.484  1000.000  1520854.00  1163000.00  0.545685E+11  1.00000
 0.95    971.500   991.235  1493373.00  1152806.00  0.475381E+11  1.00000
 1.89    929.180   970.763  1440474.00  1128998.00  0.406952E+11  1.00000
 2.84    880.000   937.635  1313516.00  1090470.00  0.342718E+11  0.76530
 3.79    880.000   908.787  1173570.00  1056919.00  0.284711E+11  0.26370
 4.73    866.030   890.879  1082536.00  1036092.12  0.232091E+11  0.00000
 5.68    839.705   871.848  1049630.00  1013959.06  0.182363E+11  0.00000
 6.63    810.895   848.553  1013616.94   986867.94  0.134242E+11  0.00000
 7.57    781.353   822.296   976690.50   956330.25  0.878216E+10  0.00000
 8.52    752.581   794.611   940724.81   924133.25  0.431018E+10  0.00000
 9.47    725.851   766.866   907312.19   891865.94  0.000000E+00  0.00000
 QK 0.545684684E+11
 QP,QG,(QG/TAUL) -677.835693      -9646193.00      -903.662841


      52.00 TAUH IN H.MIN

   X       TK        TG        UK          UG         ESP         SA
 0.00    997.981  1000.000  1526475.00  1163000.00  0.567761E+11  1.00000
 0.95    987.636   996.399  1513543.00  1158812.00  0.496856E+11  1.00000
 1.89    965.268   986.413  1485583.00  1147198.00  0.426904E+11  1.00000
 2.84    904.241   960.653  1409800.00  1117239.00  0.359372E+11  1.00000
 3.79    880.000   926.447  1226090.00  1077458.00  0.297894E+11  0.45194
 4.73    880.000   903.195  1112783.00  1050416.00  0.243343E+11  0.04582
 5.68    869.917   889.063  1087395.00  1033980.19  0.192029E+11  0.00000
 6.63    848.604   874.141  1060754.00  1016626.56  0.141927E+11  0.00000
 7.57    824.485   855.322  1030605.31   994739.31  0.931505E+10  0.00000
 8.52    798.831   833.468   998537.62   969322.94  0.458248E+10  0.00000
 9.47    773.005   809.654   966255.06   941628.19  0.000000E+00  0.00000
 QK 0.567760855E+11
 QP,QG,(QG/TAUL) -553.429931      -11836197.0      -832.971435


      53.00 TAUH IN H.MIN

   X       TK        TG        UK          UG         ESP         SA
 0.00    999.366  1000.000  1528206.00  1163000.00  0.585882E+11  1.00000
 0.95    994.711   998.517  1522387.00  1161275.00  0.514730E+11  1.00000
 1.89    983.316   993.759  1508144.00  1155742.00  0.444046E+11  1.00000
 2.84    952.974   980.936  1470216.00  1140829.00  0.374579E+11  1.00000
 3.79    880.000   948.677  1314697.00  1103311.00  0.309624E+11  0.76953
 4.73    880.000   914.292  1156808.00  1063322.00  0.251980E+11  0.20362
 5.68    880.000   897.130  1111991.00  1043362.62  0.199065E+11  0.04299
 6.63    872.637   886.712  1090795.00  1031246.31  0.147689E+11  0.00000
 7.57    855.228   875.312  1069033.00  1017988.06  0.973153E+10  0.00000
 8.52    834.948   860.167  1043684.19  1000374.06  0.480405E+10  0.00000
 9.47    812.886   842.062  1016106.69   979317.69  0.000000E+00  0.00000
 QK 0.585882132E+11
 QP,QG,(QG/TAUL) -459.206054      -13630178.0      -768.128173
```

```
        54.00 TAUH IN H.MIN

    X        TK        TG        UK          UG          ESP        SA
  0.00    999.798  1000.000  1528746.00  1163000.00  0.601213E+11  1.00000
  0.95    997.757   999.388  1526195.00  1162288.00  0.529959E+11  1.00000
  1.89    992.128   997.162  1519158.00  1159700.00  0.458930E+11  1.00000
  2.84    976.800   990.806  1499999.00  1152307.00  0.388511E+11  1.00000
  3.79    912.323   967.658  1419403.00  1125386.00  0.320419E+11  1.00000
  4.73    880.000   931.859  1216748.00  1083752.00  0.258934E+11  0.41846
  5.68    880.000   905.923  1141884.00  1053588.00  0.203923E+11  0.15013
  6.63    880.000   892.900  1107700.00  1038442.31  0.151456E+11  0.02761
  7.57    873.565   884.901  1091955.00  1029139.69  0.100153E+11  0.00000
  8.52    859.511   875.642  1074387.00  1018372.19  0.496278E+10  0.00000
  9.47    842.768   863.392  1053459.00  1004125.56  0.000000E+00  0.00000
 QK  0.601212641E+11
 QP,QG,(QG/TAUL) -397.186523       -15145125.0        -711.716064


        55.00 TAUH IN H.MIN

    X        TK        TG        UK          UG          ESP        SA
  0.00    999.934  1000.000  1528916.00  1163000.00  0.614772E+11  1.00000
  0.95    999.054   999.746  1527816.00  1162705.00  0.543477E+11  1.00000
  1.89    996.336   998.719  1524419.00  1161510.00  0.472287E+11  1.00000
  2.84    988.660   995.606  1514823.00  1157890.00  0.401399E+11  1.00000
  3.79    958.676   984.616  1477344.00  1145109.00  0.331610E+11  1.00000
  4.73    880.000   951.949  1311857.00  1107117.00  0.266555E+11  0.75935
  5.68    880.000   915.925  1189148.00  1065221.00  0.208223E+11  0.31953
  6.63    880.000   897.936  1131291.00  1044300.00  0.154103E+11  0.11216
  7.57    880.000   888.977  1105589.00  1033879.81  0.101932E+11  0.02004
  8.52    874.451   883.071  1093062.00  1027011.56  0.506528E+10  0.00000
  9.47    862.988   875.900  1078734.00  1018672.06  0.000000E+00  0.00000
 QK  0.614771589E+11
 QP,QG,(QG/TAUL) -360.820312       -16483770.0        -664.272216


        56.00 TAUH IN H.MIN

    X        TK        TG        UK          UG          ESP        SA
  0.00    999.978  1000.000  1528971.00  1163000.00  0.627409E+11  1.00000
  0.95    999.601   999.894  1528500.00  1162877.00  0.556097E+11  1.00000
  1.89    998.313   999.425  1526890.00  1162331.00  0.484833E+11  1.00000
  2.84    994.516   997.917  1522143.00  1160578.00  0.413718E+11  1.00000
  3.79    980.239   992.640  1504297.00  1154441.00  0.343129E+11  1.00000
  4.73    912.954   969.591  1420191.00  1127634.00  0.274919E+11  1.00000
  5.68    880.000   932.989  1250984.00  1085066.00  0.212617E+11  0.54117
  6.63    880.000   906.469  1162145.00  1054223.00  0.156334E+11  0.22275
  7.57    880.000   873.211  1120984.00  1038804.87  0.103085E+11  0.07522
  8.52    880.000   886.610  1102690.00  1031128.00  0.512217E+10  0.00965
  9.47    874.800   881.970  1093498.00  1025731.06  0.000000E+00  0.00000
 QK  0.627409141E+11
 QP,QG,(QG/TAUL) -343.172851       -17667312.0        -623.186767
```

```
****************************************************
DISCHARGING   TAU=   201600.
****************************************************
89


     57.00 TAUH IN H.MIN

     X      TK       TG        UK          UG          ESP        SA
   0.00   988.494  944.427  1514616.00  1069091.00  0.610069E+11  1.00000
   0.95   984.189  926.995  1509235.00  1049359.00  0.562375E+11  1.00000
   1.89   977.847  904.532  1501307.00  1023930.69  0.513160E+11  1.00000
   2.84   967.940  876.110  1488924.00   991756.37  0.461869E+11  1.00000
   3.79   948.380  841.997  1464474.00   953140.81  0.407029E+11  1.00000
   4.73   890.326  809.837  1391906.00   916736.19  0.345997E+11  1.00000
   5.68   859.054  782.873  1224821.00   886212.19  0.279375E+11  0.54124
   6.63   851.439  752.782  1126455.00   852149.81  0.210490E+11  0.22278
   7.57   841.377  713.860  1072711.00   808090.00  0.140746E+11  0.07524
   8.52   827.785  663.681  1037424.81   751287.19  0.705281E+10  0.00966
   9.47   806.017  600.000  1007520.19   679200.00  0.000000E+00  0.00000
QK-0.627412254E+11
QP,QG,(QG/TAUL)   467.868408        1735582.00        482.067138


     58.00 TAUH IN H.MIN

     X      TK       TG        UK          UG          ESP        SA
   0.00   974.628  923.130  1497283.00  1044983.50  0.593747E+11  1.00000
   0.95   966.734  903.358  1487416.00  1022601.94  0.548810E+11  1.00000
   1.89   956.108  879.216  1474134.00   995272.37  0.501868E+11  1.00000
   2.84   941.454  850.292  1455036.00   962531.12  0.452409E+11  1.00000
   3.79   918.248  817.218  1426808.00   925090.31  0.399035E+11  1.00000
   4.73   867.095  785.848  1362868.00   889580.25  0.339329E+11  1.00000
   5.68   836.796  758.367  1196999.00   858472.00  0.274262E+11  0.54124
   6.63   823.431  728.539  1091445.00   824706.50  0.207028E+11  0.22278
   7.57   806.509  692.609  1029126.44   784033.87  0.138690E+11  0.07524
   8.52   784.706  649.797   983575.69   735570.69  0.696149E+10  0.00966
   9.47   754.520  600.000   943148.44   679200.00  0.000000E+00  0.00000
QK-0.627412254E+11
QP,QG,(QG/TAUL)   438.939453        3367409.00        467.839355


     59.00 TAUH IN H.MIN
   212401.079557240009
     X      TK       TG        UK          UG          ESP        SA
   0.00   958.845  901.687  1477555.00  1020710.00  0.578471E+11  1.00000
   0.95   947.773  880.225  1463715.00   996414.44  0.535700E+11  1.00000
   1.89   933.606  855.083  1446006.00   967954.62  0.490702E+11  1.00000
   2.84   915.347  826.235  1423183.00   935297.69  0.442916E+11  1.00000
   3.79   889.743  794.510  1391178.00   899385.06  0.390968E+11  1.00000
   4.73   843.938  764.442  1333921.00   865348.19  0.332591E+11  1.00000
   5.68   814.367  737.541  1168962.00   834896.00  0.269031E+11  0.54124
   6.63   796.936  709.228  1058326.00   802846.12  0.203389E+11  0.22278
   7.57   780.000  676.298   990497.87   765569.37  0.136468E+11  0.05555
   8.52   751.095  639.205   938868.06   723579.75  0.686020E+10  0.00966
   9.47   715.972  600.000   894964.00   679200.00  0.000000E+00  0.00000
QK-0.627412254E+11
QP,QG,(QG/TAUL)   409.811523        4894428.00        453.404052
```

```
      60.00  TAUH  IN  H.MIN

     X       TK        TG         UK           UG         ESP           SA
  0.00    941.701   880.816   1456125.00    997084.37   0.564229E+11  1.00000
  0.95    927.936   858.364   1438919.00    971667.75   0.523151E+11  1.00000
  1.89    910.970   832.964   1417711.00    942915.50   0.479887E+11  1.00000
  2.84    890.133   804.867   1391665.00    911110.00   0.433686E+11  1.00000
  3.79    863.183   775.022   1357977.00    877325.19   0.383060E+11  1.00000
  4.73    821.763   746.983   1306202.00    845585.06   0.325950E+11  1.00000
  5.68    793.074   721.860   1142347.00    817145.75   0.263810E+11  0.54124
  6.63    780.000   694.980   1028295.37    786717.75   0.199678E+11  0.19103
  7.57    762.092   663.364    952613.19    750928.00   0.134152E+11  0.00000
  8.52    721.904   630.680    903379.25    713930.12   0.675239E+10  0.00000
  9.47    687.095   600.000    858867.87    679200.00   0.000000E+00  0.00000
QK-0.627412254E+11
QP,QG,(QG/TAUL)   381.460449          6317946.00        438.993164


      61.00  TAUH  IN  H.MIN

     X       TK        TG         UK           UG         ESP           SA
  0.00    923.740   861.134   1433673.00    974804.00   0.550994E+11  1.00000
  0.95    907.809   838.417   1413760.00    949088.69   0.511239E+11  1.00000
  1.89    888.771   813.527   1389963.00    920912.37   0.469437E+11  1.00000
  2.84    866.277   786.923   1361845.00    890797.06   0.424780E+11  1.00000
  3.79    838.871   759.632   1327588.00    859903.12   0.375496E+11  1.00000
  4.73    801.287   734.507   1280607.00    831461.56   0.319553E+11  1.00000
  5.68    780.000   711.177   1117931.00    805053.00   0.258719E+11  0.51230
  6.63    780.000   682.579    995128.81    722679.69   0.195991E+11  0.07215
  7.57    735.684   651.298    919563.75    737269.87   0.131807E+11  0.00000
  8.52    698.169   624.020    872709.94    706391.19   0.664134E+10  0.00000
  9.47    665.463   600.000    831827.25    679200.00   0.000000E+00  0.00000
QK-0.627412254E+11
QP,QG,(QG/TAUL)   354.724121          7640708.00        424.802002


      62.00  TAUH  IN  H.MIN

     X       TK        TG         UK           UG         ESP           SA
  0.00    905.540   842.719   1410924.00    953957.69   0.538673E+11  1.00000
  0.95    888.021   820.242   1389025.00    928513.81   0.499953E+11  1.00000
  1.89    867.650   796.298   1363561.00    901409.25   0.459366E+11  1.00000
  2.84    844.374   771.473   1334466.00    873307.94   0.416213E+11  1.00000
  3.79    817.352   746.784   1300689.00    845360.12   0.368396E+11  1.00000
  4.73    783.495   724.487   1258367.00    820119.44   0.313586E+11  1.00000
  5.68    780.000   700.689   1091573.00    793180.00   0.253899E+11  0.41783
  6.63    766.888   670.448    958608.81    758947.50   0.192436E+11  0.00000
  7.57    713.297   641.500    891619.94    726178.44   0.129507E+11  0.00000
  8.52    678.896   618.808    848619.44    700490.50   0.653060E+10  0.00000
  9.47    649.226   600.000    811531.19    679200.00   0.000000E+00  0.00000
QK-0.627412254E+11
QP,QG,(QG/TAUL)   329.708740          8872131.00        411.059570
```

```
      63.00 TAUH IN H.MIN

   X        TK        TG        UK          UG         ESP         SA
 0.00     887.416   825.756   1388268.00   934755.87  0.527228E+11  1.00000
 0.95     868.818   803.989   1365021.00   910115.56  0.489318E+11  1.00000
 1.89     847.716   781.425   1338644.00   884572.69  0.449739E+11  1.00000
 2.84     824.311   758.731   1309387.00   858884.06  0.407896E+11  1.00000
 3.79     798.246   736.889   1276806.00   834159.06  0.361562E+11  1.00000
 4.73     780.000   715.183   1236632.00   809586.81  0.307985E+11  0.93775
 5.68     780.000   688.248   1060451.00   779096.44  0.249361E+11  0.30628
 6.63     740.934   658.231    926166.81   745118.12  0.189041E+11  0.00000
 7.57     694.316   633.544    867893.69   717172.00  0.127278E+11  0.00000
 8.52     663.303   614.730    829128.12   695874.50  0.642177E+10  0.00000
 9.47     637.041   600.000    796300.31   679200.00  0.000000E+00  0.00000
QK-0.627412254E+11
QP,QG,(QG/TAUL)   306.666259        10015968.0        397.764892


      64.00 TAUH IN H.MIN

   X        TK        TG        UK          UG         ESP         SA
 0.00     869.810   810.409   1366261.00   917382.69  0.516570E+11  1.00000
 0.95     850.645   789.701   1342305.00   893941.56  0.479294E+11  1.00000
 1.89     829.392   768.781   1315739.00   870260.81  0.440548E+11  1.00000
 2.84     806.463   748.353   1287078.00   847135.69  0.399816E+11  1.00000
 3.79     781.929   729.298   1256410.00   825565.06  0.354951E+11  1.00000
 4.73     780.000   707.826   1211968.00   801259.44  0.302776E+11  0.84935
 5.68     780.000   677.837   1025022.06   767311.31  0.245204E+11  0.1/930
 6.63     718.872   648.080    898589.12   733626.12  0.185879E+11  0.00000
 7.57     678.285   627.089    847855.19   709865.19  0.125171E+11  0.00000
 8.52     650.718   611.540    813396.56   692262.81  0.631747E+10  0.00000
 9.47     627.892   600.000    784863.69   679200.00  0.000000E+00  0.00000
QK-0.627412254E+11
QP,QG,(QG/TAUL)   285.818847        11081210.0        385.062744


      65.00 TAUH IN H.MIN

   X        TK        TG        UK          UG         ESP         SA
 0.00     853.088   797.662   1345359.00   902953.75  0.506602E+11  1.00000
 0.95     833.844   778.625   1321303.00   881403.44  0.469822E+11  1.00000
 1.89     812.972   760.009   1295214.00   860330.00  0.431765E+11  1.00000
 2.84     791.086   742.542   1267856.00   840557.87  0.391970E+11  1.00000
 3.79     780.000   724.674   1237323.00   820331.12  0.348561E+11  0.94023
 4.73     780.000   701.685   1184798.00   794307.87  0.297931E+11  0.75197
 5.68     780.000   669.144    985992.44   757471.44  0.241437E+11  0.03940
 6.63     700.174   639.661    875216.44   724096.87  0.183006E+11  0.00000
 7.57     664.808   621.859    831009.56   703944.44  0.123225E+11  0.00000
 8.52     640.582   609.042    800726.87   689436.12  0.621974E+10  0.00000
 9.47     621.017   600.000    776270.69   679200.00  0.000000E+00  0.00000
QK-0.627412254E+11
QP,QG,(QG/TAUL)   268.503906        12077327.0        373.047851
```

```
          66.00 TAUH IN H.MIN
     X        TK        TG        UK            UG          ESP          SA
   0.00    837.503   785.719   1325877.00    889434.12   0.497232E+11   1.00000
   0.95    818.600   768.122   1302249.00    869514.56   0.460840E+11   1.00000
   1.89    798.542   751.309   1277176.00    850482.19   0.423349E+11   1.00000
   2.84    780.000   735.532   1251614.00    832623.00   0.384346E+11   0.99145
   3.79    780.000   717.055   1216195.00    811706.37   0.342316E+11   0.86450
   4.73    780.000   690.898   1154770.00    782096.31   0.293302E+11   0.64434
   5.68    757.359   658.569    946697.81    745500.62   0.238002E+11   0.00000
   6.63    684.320   632.687    855399.50    716201.81   0.180443E+11   0.00000
   7.57    653.525   617.626    816905.50    699152.50   0.121461E+11   0.00000
   8.52    632.435   607.088    790542.37    687223.37   0.612986E+10   0.00000
   9.47    615.849   600.000    769810.06    679200.00   0.000000E+00   0.00000
QK-0.627412254E+11
QP,QG,(QG/TAUL)   252.280365        13013767.0         361.777099


          67.00 TAUH IN H.MIN
     X        TK        TG        UK            UG          ESP          SA
   0.00    823.119   776.034   1307897.00    878470.81   0.488393E+11   1.00000
   0.95    804.882   760.253   1285101.00    860606.75   0.452305E+11   1.00000
   1.89    785.978   745.631   1261471.00    844054.87   0.415280E+11   1.00000
   2.84    780.000   730.107   1234665.00    826481.12   0.376945E+11   0.93070
   3.79    780.000   709.375   1192066.00    803012.56   0.336075E+11   0.77802
   4.73    780.000   680.028   1120324.00    769792.00   0.288631E+11   0.52088
   5.68    731.090   648.643    913860.81    734263.81   0.234697E+11   0.00000
   6.63    670.804   626.903    838504.44    709654.12   0.178096E+11   0.00000
   7.57    644.112   614.203    805138.69    695277.62   0.119876E+11   0.00000
   8.52    625.896   605.557    782369.44    685490.62   0.604793E+10   0.00000
   9.47    611.960   600.000    764949.50    679200.00   0.000000E+00   0.00000
QK-0.627412254E+11
QP,QG,(QG/TAUL)   239.124557        13897080.0         351.212646


          68.00 TAUH IN H.MIN
     X        TK        TG        UK            UG          ESP          SA
   0.00    810.149   768.476   1291685.00    869915.44   0.479992E+11   1.00000
   0.95    792.880   754.741   1270099.00    854366.69   0.444142E+11   1.00000
   1.89    780.000   741.570   1247810.00    839457.69   0.407497E+11   0.97782
   2.84    780.000   725.602   1215858.00    821381.44   0.369725E+11   0.86330
   3.79    780.000   702.999   1165427.00    795794.56   0.329823E+11   0.68254
   4.73    780.000   671.004   1082133.00    759576.44   0.283904E+11   0.38399
   5.68    709.352   640.385    886868.44    724916.62   0.231483E+11   0.00000
   6.63    659.326   622.117    824156.56    704236.62   0.175942E+11   0.00000
   7.57    636.288   611.439    795358.62    692149.19   0.118479E+11   0.00000
   8.52    620.660   604.359    775824.44    684134.44   0.597513E+10   0.00000
   9.47    609.035   600.000    761292.31    679200.00   0.000000E+00   0.00000
QK-0.627412254E+11
QP,QG,(QG/TAUL)   228.857986        14736593.0         341.413574
```

```
       69.00 TAUH IN H.MIN

   X       TK        TG        UK          UG         ESP        SA
0.00    798.871   763.003  1277587.00   363719.56  0.471892E+11  1.00000
0.95    782.919   751.412  1257648.00   850598.19  0.436238E+11  1.00000
1.89    780.000   738.925  1233520.00   836462.87  0.399895E+11  0.92880
2.84    780.000   721.858  1195543.00   817143.69  0.362596E+11  0.79048
3.79    780.000   697.700  1136670.00   789797.06  0.323503E+11  0.57947
4.73    780.000   663.504  1040786.00   751086.44  0.279074E+11  0.23580
5.68    691.329   633.516   864160.31   717140.12  0.228288E+11  0.00000
6.63    649.598   618.158   811996.75   699755.62  0.173892E+11  0.00000
7.57    629.794   609.207   787241.62   689622.19  0.117236E+11  0.00000
8.52    616.467   603.419   770582.94   683071.06  0.591321E+10  0.00000
9.47    606.829   600.000   758534.87   679200.00  0.000000E+00  0.00000
QK-0.627412254E+11
QP,QG,(QG/TAUL)   221.422927        15546029.0        332.460693


       70.00 TAUH IN H.MIN

   X       TK        TG        UK          UG         ESP        SA
0.00    789.311   759.770  1265637.00   860059.50  0.464014E+11  1.00000
0.95    780.000   749.432  1246849.00   848357.44  0.428496E+11  0.97437
1.89    780.000   736.729  1218301.00   833977.25  0.392417E+11  0.87205
2.84    780.000   718.752  1173995.00   813627.00  0.355514E+11  0.71325
3.79    780.000   693.303  1106169.00   784819.12  0.317096E+11  0.47014
4.73    780.000   657.280   996837.94   744041.12  0.274128E+11  0.07823
5.68    676.392   627.810   845489.44   710681.44  0.225078E+11  0.00000
6.63    641.385   614.892   801730.94   696058.31  0.171896E+11  0.00000
7.57    624.422   607.406   780527.06   687583.81  0.116099E+11  0.00000
8.52    613.115   602.683   766393.44   682237.75  0.586015E+10  0.00000
9.47    605.166   600.000   756455.94   679200.00  0.000000E+00  0.00000
QK-0.627412254E+11
QP,QG,(QG/TAUL)   217.030990        16333187.0        324.344238


       71.00 TAUH IN H.MIN

   X       TK        TG        UK          UG         ESP        SA
0.00    781.513   756.591  1255896.00   856460.75  0.456266E+11  1.00000
0.95    780.000   746.545  1235475.00   845089.44  0.420863E+11  0.93361
1.89    780.000   732.644  1202141.00   829353.12  0.384991E+11  0.81413
2.84    780.000   712.966  1151119.00   807077.94  0.348419E+11  0.63126
3.79    780.000   685.110  1073750.00   775544.75  0.310564E+11  0.35395
4.73    761.833   649.447   952290.06   735174.44  0.268997E+11  0.00000
5.68    663.981   623.072   829975.19   705318.06  0.221743E+11  0.00000
6.63    634.471   612.201   793088.50   693012.00  0.169851E+11  0.00000
7.57    619.989   605.955   774985.37   685941.00  0.114964E+11  0.00000
8.52    610.439   602.106   763048.31   681584.50  0.581089E+10  0.00000
9.47    603.911   600.000   754887.44   679200.00  0.000000E+00  0.00000
QK-0.627412254E+11
QP,QG,(QG/TAUL)   212.712371        17100272.0        316.937988


       72.00 TAUH IN H.MIN

   X       TK        TG        UK          UG         ESP        SA
0.00    780.000   753.757  1246841.00   853252.75  0.448688E+11  0.97435
0.95    780.000   742.850  1222730.00   840906.31  0.413375E+11  0.88793
1.89    780.000   727.415  1184099.00   823434.25  0.377672E+11  0.74946
2.84    780.000   705.564  1125580.00   798698.44  0.341375E+11  0.53972
3.79    780.000   674.635  1037330.56   763686.56  0.303994E+11  0.22341
4.73    732.206   640.776   915256.44   725358.19  0.263595E+11  0.00000
5.68    653.536   619.124   816919.00   700848.44  0.218054E+11  0.00000
6.63    628.669   609.989   785835.50   690507.19  0.167608E+11  0.00000
7.57    616.340   604.786   770424.06   684618.31  0.113737E+11  0.00000
8.52    608.306   601.654   760381.44   681072.69  0.576004E+10  0.00000
9.47    602.964   600.000   753703.75   679200.00  0.000000E+00  0.00000
QK-0.627412254E+11
QP,QG,(QG/TAUL)   208.863052        17850512.0        310.164794
```

Solar Thermal Energy Utilization
- German Studies on Technology and Application -

Index of Authors

Birke, G.:   Process Synthesis of a Gasification Process Modified for High Solar Energy Integration (Lurgi, Frankfurt), Vol. 3, p. 547.

Bitterlich, W.:   Expert Opinion and Co-operation in the Development Program High Temperature Storage Tank (Uni-Essen GHS), Vol. 2, p. 211.

Boese, F.:   Considerations and Proposals for Future Research and Development of High Temperature Solar Processes (Motor Columbus, Stuttgart), Vol. 1, p. 169.

Bohn, Th.J.:   Expert Opinion and Co-operation in the Development Program High Temperature Storage Tank, (Uni-Essen GHS), Vol. 2, p. 211.

Erdle, E.:   Utilization of Solar Energy for Hydrogen Production by High Temperature Electrolysis of Steam, (Dornier, Friedrichshafen), Vol. 3, p. 621.

Freudenstein, K.:   Volumetric Ceramic Receiver Cooled by Open Air Flow - Feasibility Study - (Interatom, Bergisch-Gladbach), Vol. 2, p. 1.

Fuhrmann, H.:   Comparative Investigations and Ratings of Different Solar Systems Using Tubular Steam Reformers, (MAN-Technologie GmbH, München), Vol. 3, p. 251.

Groß, J.:   Utilization of Solar Energy for Hydrogen Production by High Temperature Electrolysis of Steam, (Dornier, Friedrichshafen), Vol.3, p. 621.

Grychta, A.:   Literature Survey in the Field of Primary and Secondary Concentrating Solar Energy Systems Concerning the Choice and Manufacturing Process of Suitable Materials (NU-Tech-Neumünster), Vol. 1, p. 97.

Huber, P.E.:   Considerations and Proposals for Future
                        Research and Development of High Temperature
                        Solar Processes, (Motor Columbus, Stuttgart),
                        Vol. 1, p. 169.

Jäger, W.:   A Multistage Steam Reformer Utilizing Solar
                        Heat, (Interatom, Bergisch-Gladbach),
                        Vol. 2, p. 57.

Josfeld, F.J.:   Expert Opinion and Co-operation in the Devel-
                        opment Program High Temperature Storage Tank
                        (Uni-Essen GHS), Vol. 2, p. 211.

Kalfa, H.:   Layout of High Temperature Solid Heat Storages
                        (Didier, Wiesbaden), Vol. 2, p. 111.

Kalt, A.:   Solar Steam Reforming of Methane (SSRM)
                        Program Proposals, (DFVLR, Köln),
                        Vol. 3, p. 179.

Kappler, H.W.:   Considerations and Proposals for Future
                        Research and Development of High Temperature
                        Solar Processes, (Motor Columbus, Stuttgart),
                        Vol. 1, p. 169.

Karnowsky, B.:   Volumetric Ceramic Receiver Cooled by Open Air
                        Flow - Feasibility Study -
                        (Interatom, Bergisch-Gladbach), Vol. 2, p. 1.

Kaufmann, J.:   Literature Survey in the Field of Primary and
                        Secondary Concentrating Solar Energy Systems
                        Concerning the Choice and Manufacturing
                        Process of Suitable Materials
                        (NU-Tech-Neumünster), Vol. 1, p. 97.

Koepke, P.:   Yearly Yield of Solar CRS-Process Heat and
                        Temperature of Reaction
                        (Universität München), Vol. 1, p. 3.

Lammers, J.:   Considerations and Proposals for Future
                        Research and Development of High Temperature
                        Solar Processes, (Motor Columbus, Stuttgart)
                        Vol. 1, p. 169.

Lensch, G.:   Literature Survey in the Field of Primary and
                        Secondary Concentrating Solar Energy Systems
                        Concerning the Choice and Manufacturing
                        Process of Suitable Materials
                        (NU-Tech-Neumünster), Vol. 1, p. 97.

Leuchs, U.: Solar Steam Reforming of Methane - Program Proposals, Vol. 3, p. 195; A Multistage Steam Reformer Utilizing Solar Heat, Vol. 2, p. 57, (Interatom, Bergisch-Gladbach,

Lippert, P.: Literature Survey in the Field of Primary and Secondary Concentrating Solar Energy Systems Concerning the Choice and Manufacturing Process of Suitable Materials (NU-Tech-Neumünster), Vol. 1, p. 97.

Meyringer, V.: Utilization of Solar Energy for Hydrogen Production by High Temperature Electrolysis of Steam, (Dornier, Friedrichshafen), Vol. 3, p. 621.

Müller, W.D.: Steam Reforming of Methane Utilizing Solar Heat, Vol. 3, p. 1; Comparative Investigations and Ratings of Different Solar Systems Using Tubular Steam Reformers (Lurgi, Frankfurt), Vol. 3, p. 251.

Quenzel, H.: Yearly Yield of Solar CRS-Process Heat and Temperature of Reaction, (Universität München) Vol. 1, p. 3.

Reimert, R.: Process Synthesis of a Gasification Process Modified for High Solar Energy Integration (Lurgi, Frankfurt), Vol. 3, p. 547.

Siebert, W.: A Multistage Steam Reformer Utilizing Solar Heat, (Interatom, Bergisch-Gladbach), Vol. 2, p. 57.

Sizmann, R.: Yearly Yield of Solar CRS-Process Heat and Temperature of Reaction, (Universität München) Vol. 1, p. 3.

Streuber, Chr.: Layout of High Temperature Solid Heat Storages (Didier, Wiesbaden), Vol. 2, p. 111.

Werner, K.: Expert Opinion and Co-operation in the Development Program High Temperature Storage Tank (Uni-Essen GHS), Vol. 2, p. 211.